Science as a Spiritual Practice

Science as a Spiritual Practice

Imants Barušs

imprint-academic.com

Copyright © Imants Barušs, 2007

The moral rights of the author have been asserted.
No part of any contribution may be reproduced in any form
without permission, except for the quotation of brief passages
in criticism and discussion.

Published in the UK by
Imprint Academic, PO Box 200, Exeter EX5 5YX, UK

Published in the USA by
Imprint Academic, Philosophy Documentation Center
PO Box 7147, Charlottesville, VA 22906-7147, USA

ISBN 978 184540 0743

A CIP catalogue record for this book is available from the
British Library and US Library of Congress

Contents

Acknowledgements	vi
Prologue: Forbidden Research	1
Part I: Beyond Materialism	5
Part II: Access to Inner Knowledge	47
Part III: Seeking Transcendence	91
Epilogue: Science as a Spiritual Practice	125
Endnotes	127
Recommended Reading	131
References	133
Index	147

Acknowledgements

I thank the following scholars for the effort that they put into reading all or parts of the manuscript and providing me with feedback: Gerry McKeon, Roger Nelson, Lynne Jackson, Sauro Camiletti, Allan Combs, Ron Leonard, Charles Tart, Sandra Sebre, Thomas McFarlane, and Ian Brown. None of the readers is responsible for any errors that may yet remain in the book, nor do they necessarily endorse any of the ideas that I express. I am grateful to Mark Debicki for giving me permission to recount his experiences of trying to induce lucid dreaming. I thank Allan Combs and the editors of the journal *Integralis: Journal of Integral Consciousness, Culture, and Science* for permission to reproduce parts of my paper "Transition to Transcendence: Franklin Merrell-Wolff's Mathematical Yoga" that was published in their journal. I appreciate the conscientious help provided by my research assistants, Allison Foskett, Shannon Foskett, Elizabeth Russell, and Robert Horvath. In particular, I am grateful for the comprehensive comments about the manuscript and editing assistance provided by Shannon. I want to thank Keith Sutherland and Anthony Freeman of Imprint Academic for their enthusiasm for my work, for their co-operative spirit, and for the quality with which they have published this book. Finally, I thank King's University College at The University of Western Ontario for research funds that were used to support this project.

How else can one write but of those things which one doesn't know, or knows badly? It is precisely there that we imagine having something to say. We write only at the frontiers of our knowledge, at the border which separates our knowledge from our ignorance and transforms the one into the other . . . We are therefore well aware, unfortunately, that we have spoken about science in a manner which is not scientific. — *Gilles Deleuze*

Prologue
Forbidden Research

I have written this book as a result of a comment made at a scientific meeting. One of the participants, having become persuaded of the value of spiritual aspiration, asked that a spiritual practice be developed that would be suitable for scientists. What is little known is that such a practice already exists, namely, that of "mathematical yoga", originated by Franklin Wolff. The idea is that scientists who become interested in spirituality do not need to abandon their scientific expertise and start afresh with some other sorts of activities, but that what is required is a reorientation of the skills that a scientist has already developed so that science itself can become a spiritual practice.

Scientists are generally not supposed to be interested in spirituality except for the study of the manner in which such incidental interests can arise in a biological organism and play themselves out in our interactions with one another. As an example of the type of research that is being done, in a recent experiment using brain imaging techniques, it was found that acceptance of spirituality is related to the functioning of the serotonin system of the brain.[1] That is interesting to know, but of marginal significance when trying to answer existential questions as spiritual aspirants attempt to do. To seriously investigate fundamental questions about reality without the constraints of accepted doctrines in science is tacitly forbidden. Not only is support for spiritual aspiration lacking within the scientific community, but such research interests can be met with vociferous derision. Spiritual exploration is forbidden research.

This book is written for those who want to explore the interface between science and spirituality in an open-minded manner. My purpose is not to try to change the opinions of those who are convinced that such exploration is blasphemous. This is an important distinction. Challenging true believers of the predominant paradigm in science would require a different approach — a more pedantic style consisting of careful arguments against the current paradigm's premises. Rather than writing in that fashion, I am presupposing that readers are ready to think outside the confines of mainstream science and so I am writing about a broader range of matters in a more conversational style. This leaves my exposition vulnerable to the castigation that typically gets levelled at this type of subject matter. However, I feel that any such criticism is a necessary price to pay for the sake of having a serious discussion about science and spirituality.

Scientists have unavoidably been steeped in conventional science and, in particular, the materialist ideology that underlies much of it. Hence that is where we pick up the thread in Part I. From there, the discussion leads into the notions of spiritual aspiration, self-transformation, and the possible existence of transcendent dimensions of reality. Perhaps the most ubiquitous feature of spiritual practices is the emphasis on obtaining guidance from within oneself and so, in Part II, we examine this contention and the possibility of access to inner knowledge. But spiritual aspiration also holds out the promise of actual enlightenment. It is in this context that Franklin Wolff's philosophy is introduced in Part III as part of an explication of a scientific approach to transcendent states of consciousness.

In this book I have retained some "religious" terminology in presenting the material rather than having adopted strictly "scientific" language. I have done so in part because the development of my own understanding of transcendent states of consciousness has been largely within the "spiritual" context. And I have done so also because Wolff has relied on mixed terminology (Merrell-Wolff, 1995b).

I am a private person, but the nature of the subject matter and an effort to convey my own understanding of it has meant that I have shared some of my experiences in this book. Readers of earlier drafts have been divided regarding the inclusion of my personal experiences. Some have said that the book was too personal and others have said that it was too scientific. I have decided to stay with a hybrid version without getting too scientific or too personal. Readers who would like to see more science can consult the endnotes, in which I have included additional technical information, and the references to other sources. There is also a list of recommended reading after the Epilogue. Readers who would like more personal accounts, well, they will just have to draw on their own experiences for now. I hope that all readers of this book enjoy it and find it helpful for their own journeys of discovery.

Part I
Beyond Materialism

In the year 2000 I hosted the annual meeting of the Society for Scientific Exploration. The membership of the society consists of scientists who are interested in anomalous phenomena. The meetings last for a few days and usually include an excursion. London, Ontario, where I live, is not far from Lake Huron and I regularly go to a public park on the eastern shore to watch the sunsets across the water, which are reputed to be among the most beautiful in the world. What better excursion, I thought, than a trip to the park to watch the sunset. And so we went.

June 9, 2000 turned out to be a beautiful evening with a clear sky. As the sun descended toward the horizon, the fifteen or so conference participants who were on the beach became still and gazed silently at the setting sun. Eventually, after the sun had set, we became animated once again, and left the park.

Roger Nelson came on that excursion and ran a portable random event generator (REG) interfaced with a palmtop computer. The REG used by Nelson was a micro-electronic noise source whose signals are configured so as to yield a random string of zeros and ones. Two hundred of these zeros and ones get added together every second giving a number with an expected mean of 100. The differences between the actual and expected mean values are summed over time so as to produce a cumulative deviation from expectation. The resultant measure is an indication of the degree to which the REG is departing from random behaviour.

A REG, such as the one used by Nelson, is designed to function randomly, so that only chance deviations from expected values should occur. However, when Nelson showed me a graphical display of the statistical analyses of the data from 30 minutes prior to the time of the sunset until 30 minutes after the sun had set, there was a striking deviation of the REG from random behaviour. Indeed, when the one-hour interval bracketing the sunset is considered as a pre-specified data segment, the probability that such a deviation occurred by chance is only 8 in 1000, well below the probability of 50 in 1000 that is usually required for establishing an event as a scientific fact. In other words, there was synchrony that was not mediated by any known physical mechanisms between the entranced activity of a group of people watching a sunset and a randomly functioning mechanical device. Indeed, when REGs have been taken to various events, it has been found that deviations from expected random behaviour can occur, particularly when there is subjective resonance among participants, as appeared to have been the case among those of us who watched the sunset (Nelson, 2000; Nelson, Jahn, Dunne, Dobyns, & Bradish, 1998).

Deconstructing Matter

Materialism is dead. However, the bones of materialism continue to rattle along the corridors of academia. To say that materialism is dead is either an outrageous or banal remark depending upon one's point of view. We will get to that disparity of views in a bit. Let us first consider additional grounds for the contention that materialism is no longer a useful interpretation of reality.

What is materialism? As a concept and, as is the case with many abstract concepts, materialism does not have clear-cut boundaries. In general, it signifies a world view that regards matter as ontologically primitive. In other words, whatever goes on in reality is a manifestation of physical processes. Let us use the term "physicalism" to refer to the idea that reality is entirely physical in nature. In practice, the notion of materialism appears to include more than what is entailed in physicalism. It appears to include the implicit cognitive schema that all of reality is ultimately made up of very small particles that collide with one another in a manner analogous to the collisions of billiard balls on a billiard table. That is to say, reality, whatever aspect of it we may be talking about, is nothing other than a series of predetermined interactions of very small particles. Given that cognitive schemata serve to structure our experience, materialism provides us with a way of thinking about the world.

What is wrong with such a world view? Well, materialism cannot readily account for the synchrony that occurred at the sunset described above. Nor can it deal with existential issues. After all, how is it that anything exists in the first place? But there is an even more lethal problem: materialism cannot account for the properties of matter. Simply put, in spite of our naive suppositions about its nature, matter, on a small enough scale, does not behave like very small billiard balls.

The inadequacy of materialism as an explanation of matter is not a recent revelation. The story of twentieth century physics was that of deconstructing both theoretically and experimentally the particulate nature of matter. Matter can resemble particles in some circumstances, but that is not its universal character. The applicable theory has been quantum mechanics and its various derivations such as quantum field theory. And the experimental evidence has resulted from testing some of the more striking predictions of quantum mechanics.

There has been a number of popular books written about the ontological implications of quantum mechanics, some of which extrapolate further from the evidence than others. We will entertain some of the more fanciful ideas later, but the deconstruction of matter is an immediate consequence of both the theory and particular experimental results, and is not the end-point of an intricate web of reasoning. Let us consider two examples of quantum phenomena that illustrate this.

Quantum Weirdness

One of the violations of materialist assumptions about matter occurs with the persistent correlations of subatomic events that originated from the same source but have subsequently become separated in space. Consider, in particular, the creation of a pair of photons with the same polarization that move away from one another at the speed of light. If one of the photons were to encounter a Polaroid filter set in the same direction of polarization as the photon, then the photon would pass through the filter. If a filter were set at right angles to the direction of polarization of the photon, then it would be blocked. But what about angles that are in between? Well, if a stream of such polarized photons were sent toward a filter at an intermediate angle, then some of the photons would pass through and some would be blocked. Thus, a single photon would either pass through or be blocked. Now, if a second filter, set at the exact same angle as that of the first filter, is subsequently placed in the path of the second photon of the pair travelling in the opposite direction to that of the first, the second photon will behave in the same manner as the first photon. That is to say, if the first photon passed through the filter set in its path, then the second photon will pass through its filter. If the first was blocked, then the second will be blocked.

Not so strange at all, you might say. At the time that the photons were created, they each got a plan that would tell them what to do were they to encounter filters set at various angles. What looks like purely probabilistic but synchronous behaviour is, in fact, governed by hidden variables. The photons are just carrying with them the information about what to do in different circumstances.

If it were the case that there were a plan determining what the photons were to do upon encountering various situations, then, based just on probability and logic, particular relationships must hold among the numbers of photons passing through filters set at various angles. These relationships are given by Bell's Theorem.[2] However, the numbers of photons that can pass through Polaroid filters set at angles to one another is theoretically given by Malus' cosine-squared law. The problem is that Malus' cosine-squared law conflicts with Bell's Theorem.[3] In other words, for some settings of the filters, Bell's Theorem would be violated. Hence the photons could not have been carrying the information with them.

What do the actual experiments reveal? Variations on this protocol have, for the most part, confirmed violations of variations on Bell's Theorem, although there has still been considerable controversy about the interpretation of the results (Aspect, Dalibard, & Roger, 1982; Wigner, 1976/1983b; Wick, 1995). On balance, it seems to me that there has been sufficient support for predictions that are consistent with quantum mechanics to seriously challenge materialist interpretations of reality.

But how can something that should be as obvious as Bell's Theorem not be correct? What does that tell us about the nature of reality? The evidence suggests that photons cannot be carrying information with them about what to do when they encounter Polaroid filters set at various angles. In other words, when one of the pair of photons in the preceding protocol encounters a Polaroid filter set at an angle to its direction of polarization then, in a probabilistic manner, it either passes through the filter or is blocked. The other photon, travelling in the opposite direction, somehow knows the angle of the Polaroid filter encountered by the first photon and, on that basis, determines, again in a probabilistic manner, but taking into account the angle of the first filter, whether or not to pass through its Polaroid filter.

Is it possible that information about the angle of the filter encountered by the first photon is somehow relayed to the second photon? The photons are moving away from one another at the speed of light, considered to be the maximum speed at which anything can travel in the universe. In the more refined experiments to test Bell's Theorem, the angles of the filters are changed too quickly for the information about the orientation of the filter encountered by the first photon to travel at the speed of light to the second photon. In other words, it is not possible for the second photon to acquire information about the behaviour of the first photon through any known means. The point is that the action of the second photon depends on what happens to the first photon irrespective of the distance separating them, indicating an inherent non-local connectedness in the universe. Such non-locality, although it does not explain the synchrony that occurred at the time of the sunset as described earlier, is at least consistent with it.

The second example that I want to use, that of the double slit experiment, is perhaps the prototypical illustration from quantum mechanics of the breakdown of our ordinary notions about reality. Suppose that we have a device that can send electrons, one at a time, toward a barrier with two slits in it. On the side of the barrier opposite the electron gun is a screen that can detect the impacts of individual electrons. Suppose that we were to close one of the slits so that only one slit were to remain open. What we would see if we were to fire electrons at the barrier would be that the electrons would land on the detecting screen opposite the open slit. Well, the electrons do not all land exactly on one spot. Rather there would be a normal probability distribution associated with where they would land, with the greatest number being detected directly across from the slit and far lesser numbers away from the centre. The same would happen, of course, if the other slit had been opened instead.

Now suppose that both slits were to be open at the same time. Our intuition tells us that, since the electron is a particle, it should either hit the barrier or go through one or the other of the two open slits. If the electron were to go through one of the slits, we should find it across from that slit. Hence, the probability distribution with both slits open should just be a normalized sum of the probability distributions for each of the slits being open on its own.

What happens, however, is that the locations at which the electron is found are spread across the detecting screen. Instead of the simple sum of normal probability distributions that we would expect, we get a complex probability distribution that indicates the presence of an interference pattern for waves. It would appear that the electron went through both slits at once and then interfered with itself as a wave before getting to the detecting screen.

How does that make sense? Surely the electron went through one or the other of the open slits. We can test for that. We can place electron detectors at each of the two slits so as to determine through which of the slits the electron emerged on its way to the detecting screen. So, we turn on the electron gun, leave both slits open, and turn on the electron detectors. When we do that we are back to the simple sum of probability distributions seen in the single slit case. We know through which slit the electron passed, and it clearly struck the detecting screen more or less directly across from that slit.

But this result does not improve the situation. Now the behaviour of the electrons appears to depend upon our decision whether or not to ascertain through which of the slits they travelled. If we turn on the electron detectors, then we get a simple sum of individual probabilities, and if we turn them off, then we get an interference pattern. When we watch, then the electron behaves as a particle; when we do not watch, then the electron behaves as a wave.

Perhaps we can escape from this conundrum. What if we wait to make the decision of whether to turn on the electron detectors until after an electron has passed through the slits? If we were to carry out such a delayed-choice experiment, then we would find the same results as we would have without the delay. Somehow the electron knows when it arrives at the barrier, whether or not a decision will be made later to detect through which of the slits it has gone. When the decision is made later to turn on the electron detectors, then we see a simple sum of normal probability distributions, and when the decision is made later not to turn on the electron detectors, then we see an interference pattern on the detecting screen. Because a later decision affects an earlier event, we end up with backward causation in time.

Various experiments have been done, not always in the form that I have described, which have confirmed the characteristics of particles as given in this example. Electrons do not behave like billiard balls, deterministic sequences of physical actions do not always occur, and observers of physical events can dramatically alter the course of those events. Thus, our ordinary intuitions about matter are incorrect.[4]

Anomalous Phenomena

A critic could say that perhaps materialism is wrong, given that it depends upon simplistic schemata about the nature of matter, but that surely physicalism must be correct. After all, matter is still matter even if somewhat different in its behaviour from what we might have naively expected. Everything must be physical whatever the physical universe turns out to be like. The problem with that argument is that what the universe turns out to be like might stretch the notion of "physical" to the point of rendering it meaningless. Sir James Jeans has said: "The universe can be best pictured, although still very imperfectly and inadequately, as consisting of pure thought, the thought of what, for want of a wider word, we must describe as a mathematical thinker" (Jeans, 1930/1937, p. 168). It is not just that matter does not behave in a matter-like way, but there exist anomalous phenomena whose occurrence suggests that the universe contains a robust dose of consciousness that promises to strain any physicalist theorizing. Let us consider several examples.

When presented with unexpected stimuli, such as emotionally provocative pictures, human beings respond with a distinctive pattern of physiological activation. Various measures can be taken to detect such activation, including measures of the electrical conductance of the skin. In one experiment, 24 participants were each shown a series of photographs while their skin conductance was measured. Some of the photographs were expected to have a calming effect and others of them were expected to have an arousing effect. As anticipated, afterwards there were increases in skin conductance for the emotionally arousing but not the calming photographs.

What is surprising, perhaps, is that the levels of skin conductance for emotionally arousing photographs were higher than those of calming photographs for about 3 seconds prior to the presentation of the photographs. Furthermore, using data from 33 participants, when emotionally arousing photographs were separated into those with erotic themes and those with violent themes, it was found that there was greater prior skin conductance for the photographs with erotic themes than those with violent themes, suggesting that participants' bodies were reacting to the meaning and not just the shock value of the pictures. When the anomalous presentiment study was replicated in a separate laboratory by a different researcher with 16 participants, again, the levels of skin conductance were higher prior to viewing emotionally arousing photographs than calming photographs, and photographs with erotic rather than violent themes (Radin, 1997a; 1997b; 2004; Bierman & Radin, 1999).

We have already considered the synchrony of a single REG with the apparently entranced activity of a group of people watching a sunset. Nelson, together with some of his colleagues, has set up a network around the world of over 40 REGs coupled to computers whose output is fed back to a central location in Princeton, New Jersey. The outputs are used to look for any deviations from randomness or intercorrelations of the REGs. The behaviour of the REGs should not only be random, but also independent of one another.

What researchers have found, however, is that the REGs show more structured behaviour than expected around the time of world events that draw the attention of the public. For example, on September 11, 2001, two airliners were maliciously flown into the twin towers of the World Trade Center in New York City, another airliner was flown into the Pentagon, and a fourth crashed in a field in Pennsylvania. These events were immediately publicized in the news media and they galvanized public attention. About two hours after the beginning of the tragic events of September 11 and lasting for several days, the data from the REGs were more structured than expected indicating correlated behaviour that could not be explained through known possible effects. In fact, in an independent analysis of the outputs from the REGs, it was found that the largest daily average correlation for the year 2001 occurred on September 11. Although this research has been contested, it would appear that deviations from expected random behaviour of REGs can occur in synchrony with meaningful human events (R. Nelson, 2002; 2003; Radin, 2002; Scargle, 2002).

The Politics of Science

I have argued that materialism is wrong on the basis of evidence about the nature of matter and the occurrence of anomalous phenomena. Matter does not comply with our naive intuitions about its nature. Presentience challenges our notions about determinism. And the synchronistic behaviour of REGs is contrary to the mechanistic conduct of billiard balls. For a materialist interpretation of the world to be false, only one counterexample is necessary. We have already seen five and there are many others (other counterexamples can be found in Radin, 1997a; Cardeña, Lynn, & Krippner, 2000; and the *Journal of Scientific Exploration*). Why, then, the pervasive persistence of materialism?

I think that there are a number of reasons. The first is that academics, including many scientists, appear to be unaware of advances in science outside their own areas of expertise; and, if they are aware of them, have not really thought about the implications of such advances. There has been some awareness of more salacious anomalous phenomena, such as the anomalous resonance between human activity and REGs, but a lack of awareness of the robustness of the scientific evidence supporting the existence of such phenomena (Radin, 1997a). This is compounded by the problem that sometimes studies of anomalous phenomena have been incorrectly described in secondary sources leaving readers with the erroneous impression that nothing anomalous has ever really happened (Child, 1985). Quite frankly, though, most academics neither have the time, nor perhaps the expertise, to read the primary literature, nor to seek out and witness for themselves the empirical investigations on which judgments about anomalous phenomena are based. This is true even more so of the interested general reader whose access to the necessary resources could be restricted. The result is that many people end up trusting putative experts for an evaluation of the evidence.

The problem with listening to the putative experts is that there is a strong bias in the scientific community against research that challenges a materialist view of reality. If anything, that is an understatement. The politics of science are such that open discussion of fundamental problems with materialism is strongly discouraged. There is a power structure in science, held in place through the admission of students to graduate programs, the approval of their research projects and theses, the hiring and promotion of university faculty, the approval of research grants, and the review and acceptance of academic papers and books for publication. The power of academics and scientists permits them to exclude from the scientific community half-baked fantasies about reality. The problem is that materialism is so entrenched that any disapproval of it can also be censored. There is a price to pay for challenging authority and that price is exclusion from the academic community. And thus, prevailing doctrines, such as that of materialism, can be sustained despite the evidence against them (cf. Kellehear, 1996; Jahn, 2001).

If all of this sounds as though something has gone terribly wrong, then I agree that, yes, it has. Science, as I understand it, was intended to be an open-ended exploration of nature and not a foreclosure of lines of investigation that tread upon prevailing opinions. Science in its oppressive form has sometimes been called "scientism" and is not authentic science at all. For scientism, a materialist world view is dominant and governs what interpretations of the world are to be acceptable. For authentic science, world views are shaped by the actual evidence that emerges in the course of examining whatever subject matter is of interest using whatever methods seem most appropriate. Furthermore, scientism has as its goal the collection of an objective body of facts, whereas authentic science aims at true understanding of reality. In other words, I think that the point of science is the acquisition of actual knowledge by a scientist. And, of course, such knowledge could be contrary to popular opinions, including materialism (Barušs, 1996).

Material versus Transcendent Beliefs

At the root of the persistence of materialism, I think, is the fact that scientists are just people and, as such, they have their own beliefs about reality. Such beliefs are not supposed to influence a scientist's research activity. However, whereas a scientist would expend considerable effort to master a body of knowledge and specific skills necessary to work in her field, those skills do not usually include the development of psychological self-mastery so as to be able to set aside her own opinions in order to clearly understand that knowledge. Of course, in many areas of science, one's private beliefs may not substantially interfere with one's scientific work. However, that is not true in all cases. In particular, for example, scientific and, more broadly, academic research concerning consciousness has been strongly correlated with individual investigators' beliefs about reality as we shall see in this section. Scientists' beliefs can interfere with their ability to follow the evidence.

Robert Moore and I investigated the role of personal beliefs in the study of consciousness. In 1986 we conducted a survey of academics and professionals who could potentially write about consciousness in the academic literature in order to determine the relationship between their ideas about consciousness and their beliefs about reality. The 334 completed questionnaires were statistically analysed in such a way that the contours of the respondents' world views were revealed. We found that these world views fell along a material versus transcendent dimension that had three main positions (Barušs, 1990a).

At the material pole of the material-transcendent dimension are beliefs that only that which is physical is real, that science is the proper way to know anything, and that everything functions in a deterministic manner. Materialists are physical monists in that they believe that only that which is physical is real. For materialists, as revealed in our study, consciousness is an emergent property of neurological or computational processes, is always about something, and is not really all that important.

There is a midway point along the material-transcendent dimension characterized by a dualist view that reality consists of both physical and non-physical aspects. What non-physical aspects could these be? In our analyses, Moore and I found two: the presence of meaning and the importance of spirituality. Meaning is something that is not inherently physical. To use the language of logic, there is a semantic richness to existence that is not exhausted by its formal syntax. Indeed, it is consciousness that is perceived to give meaning to reality. But respondents to our survey who tended toward this dualist position also had what could be regarded as religious beliefs, such as belief in an original creator and belief in the continuation of personal consciousness after death. Consciousness is important because consciousness provides evidence of the existence of a spiritual dimension within people. Unlike the focus on objectivity of the materialist position, subjective features of consciousness are emphasized in this conservatively transcendent position so that consciousness is conceptualized as a stream of thoughts, images, and feelings that go on for a person within her experience, or consciousness is thought of as the sense of existence that a person can have for herself.

Whereas Moore and I had expected to find these two world views, we were surprised to find the existence of a third, extremely transcendent position that is marked by three aspects. The first aspect consists of respondents' beliefs that they have had experiences that science could not explain, such as transcendent or out-of-body experiences, and that their ideas about life have changed dramatically in the past. The second aspect is that of extraordinary beliefs, including beliefs in the primacy of consciousness, extrasensory perception, and the existence of latent modes of understanding that are superior to rational thought. Those tending toward this position are mental monists in that consciousness is regarded as the ultimate reality and the physical world is viewed as a by-product of consciousness. The third aspect is that of inner growth, consisting of self-examination and self-transformation, enabling one to fully experience consciousness. For those tending toward the extraordinarily transcendent position, some form of consciousness is the ultimate reality and hence consciousness is supremely important.

In a way, materialism reflects faith in the veracity of our naive sensory experience. As such, materialism can function as a starting point for contemplation of reality. Of course, for some of us, our initial position from childhood may have been a religious one, but more about that in a while. The point is that our ordinary everyday encounters with the world generate inferences about its nature such as that of materialism. But we know from physics that the physical world does not comply with such simplistic attributions. So what is the relationship between our apparently self-evident conclusions about reality based on our everyday perceptions and the actual nature of reality? What bearing do our sensory impressions have on the truth? How can we know what is real? These seem like innocent questions, but by persistently asking them it may be possible to deconstruct our ordinary ideas based on sensory experience and break through the superficial layers of our interpretation of the world.

To understand the problems with materialism requires considerable sagacity and yet, ironically, those who believe that there is more to reality than meets the senses are thought to lack critical judgment (cf. Tobacyk & Milford, 1983). Indeed, not only can transcendent beliefs be considered to be grounds for a lack of critical judgment, but they can be deemed to be a sign of a mental disorder. One of the symptoms of schizotypal personality disorder is said to be the presence of magical thinking and unusual beliefs (American Psychiatric Association, 2000). And while some such ideation, such as superstitiousness, can rightly be regarded as dysfunctional, unconventional thinking in itself, such as belief in telepathy, for which there actually is some empirical evidence, should not be considered pathological. Those with transcendent beliefs are neither necessarily stupid nor mentally ill.

One of my students, Sonya Jewkes, empirically examined the personality correlates of beliefs about consciousness and reality among 75 undergraduate university students. Her most striking finding was that transcendent beliefs are correlated with a personality trait known as "understanding". Indeed, almost one third of the variation in beliefs about consciousness and reality can be attributed to this trait. This means that those who are more curious about the world, who have a more rational approach, who are more reflective, are also those with more transcendent beliefs. Furthermore, those tending toward the extraordinarily transcendent position are less likely to seek approval from others or to care what others think of them (Jewkes & Barušs, 2000; the descriptions of the personality traits are taken from Jackson, 1999).

Given that an intellectual approach to the world is correlated with transcendent beliefs, could it be that those with such beliefs are actually more, rather than less, intelligent than those with materialist beliefs? Another of my students, Nicole Lukey, followed up on Jewkes' research by giving 39 undergraduate psychology students intelligence tests as well as a measure of their beliefs. Although she did not find an overall statistically significant linear correlation between intelligence and transcendent beliefs, there were numerous instructive patterns in the data. For instance, there was a strong linear correlation between IQ and the belief that there is more to reality than the physical universe. There was also a linear correlation between IQ and the contention that one has had experiences that science would have difficulty explaining. Much of the data had a somewhat parabolic shape, in that those with lower and higher IQ scores tended to have more transcendent beliefs whereas those with midrange scores had more materialist beliefs. This was most pronounced for Performance IQ which is a measure of a person's ability to utilize memory, capacity for visualization, rapidity of mental manipulation, critical judgment, and motivation to adapt to novel situations.[5] The parabolic pattern in the data suggests that there could be different reasons for transcendent beliefs. Those with lower IQ could be uncritically reflecting transcendent ideas that are prevalent in popular culture whereas those with higher IQ could have reasoned that transcendence is more sensible than materialism. While this study is not definitive, at least for an undergraduate student sample, some suggestive correspondences between intelligence and transcendent beliefs were found (Lukey & Barušs, 2005).

The Compelling Quality of Personal Experience

We started out by considering examples of quantum weirdness, indicating that materialism cannot be correct. Not only can materialism not be correct, but the actual nature of matter and the scientific evidence for other anomalous phenomena, such as the anomalous resonance described previously, suggest that physicalism is unlikely to work as an explanation either. Or at least, given that consciousness in some form plays a role in nature, it is not clear what is gained by insisting that everything is ultimately physical. That is to say, if reality inherently has mind-like qualities, then why continue to call them physical? The reasonable course of action would be to leave the ontological questions open until there is sufficient evidence to settle them, but the politics of science are such that materialism prevails in spite of the evidence against it. Scientists are people with personal beliefs organized into social structures within which their beliefs are perpetuated so that overt movement away from materialist toward transcendent beliefs is professionally precarious.

The shift from materialist to transcendent beliefs that I have tried to indicate so far in this book is a logically rigorous one in which I have questioned the fundamental nature of reality in light of scientific research. But that is not how most people arrive at transcendent beliefs. Most such people believe that they have had experiences that science cannot explain and hence that there must be more to reality than the physical universe. This can be judged both by reading case studies in the academic literature and by noting that Moore and I found that those who believed that they had had unusual experiences also reported that their ideas about life had changed dramatically in the past (Baruss, 1990a). Let me briefly describe some situations in which shifts toward transcendent beliefs have occurred.

The first of those situations is that of near-death experiences (NDEs) whereby people report events that ostensibly occurred while they were close to death. These events can include feelings of peace, seeing one's own body and the events surrounding it, going through a tunnel toward a loving light, encountering deceased relatives, experiencing a review of one's life, and re-entering one's body. Sometimes these experiences are reported as having the quality of being more real than ordinary reality. They can also include perceptual abilities that are not normally available to a person. For example, some people who have been blind from birth and who have had NDEs have reported sight during the experiences, something that does not occur even in their dreams (Ring & Cooper, 1997). Whatever it is that happens during NDEs often has the effect of changing a person's beliefs about the nature of reality. In general, there is a reduction of anxiety about death among experiencers, and, in some cases, experiencers feel that they have been where you go when you die and come to believe that death is not the end of life (Baruss, 2003; Greyson, 2000; Fenwick & Fenwick, 1995; Ring & Valarino, 1998). Clearly this is a movement away from materialism toward transcendent beliefs.

A second situation in which a shift toward transcendence can occur is that of drug-induced states of consciousness. Ayahuasca is a psychedelic brew consumed in the upper Amazon region typically made up of *Banisteriopsis* vines and *Psychotria viridis* leaves, which contain the psychoactive beta-carbolines and *N,N-dimethyltryptamine* (DMT), respectively. Ingesting the brew has the effect of changing the drinker's frame of mind so that the world is seen as being inherently invested with meaning (Shanon, 2002). The use of the term "meaning" here is not just in the sense of providing referents for the words of a language, but in the sense of satisfying existential yearnings. For a materialist, the world with everything in it, including human beings and consciousness, is a meaningless mechanical contrivance. Meaning, as are consciousness, will, and other human attributes, is simply an emergent property of physical processes. Intoxication with psychedelics challenges that notion with direct experiences of the apparent meaningfulness of reality.

It is not just that ayahuasca creates a sense of meaningfulness, but psychedelic drugs can give the intoxicated the sense that they have encountered aspects of reality that are normally invisible to them. In the most controversial experiment in the history of the psychology of religion, on April 20, 1962, ten mostly theological students were each given 30 milligrams of psilocybin prior to a Good Friday service. One of those participants has subsequently revealed his identity and has said that as a result of his psychedelic experience, his religious understanding has become grounded in something deeper than just intellectual theory and belief (Malmgren, 1994; Barušs, 2003; Hood, Spilka, Hunsberger, & Gorsuch, 1996; Pahnke, 1963; Doblin, 1991). In an even more extreme sense, healthy participants, in a study in which they were injected with DMT in a hospital setting, felt that they had been propelled into another reality. They reported seeing cellular activity, rooms made especially for them, and alien landscapes. But that was not all. They reported encounters, often terrifying, with insectoid, reptilian, clown-like, humanoid and other alien beings. Some of the participants emphasized that these were not hallucinations produced by intoxication, but lucid observations of an objective reality (Strassman, 2001). These are not just recent observations. William James, writing about the effects of nitrous oxide, has said that he has been forced to conclude that there exist forms of consciousness of which we are ordinarily unaware but which need to be taken into account for a complete description of the universe, with the implication that these forms of consciousness have the same ontological status as ordinary waking consciousness (James, 1902/1958; see James, 1904a & 1904b for a discussion of the nature of ordinary experience).

The third situation in which people can come to adopt transcendent beliefs about reality is that in which they report that they have been abducted by aliens. The characteristic features of these experiences include intense fear, the perception of unidentified flying objects and humanoid aliens, lost time, memories of being subjected to physically invasive procedures, and apparently inexplicable physical lacerations (Appelle, Lynn, & Newman, 2000; Jacobs, 2000; Mack 1994). In some cases, events have occurred that have led experiencers to believe that they really have encountered aliens. For example, one experiencer told me about events that occurred one night when she was a college student. She had decided to sleep on the floor of her bedroom in an effort not to be taken when she awoke to see a grey beam of light move quickly across her bed. It moved across again, more slowly, and again, more slowly still. Then, suddenly, the light receded but the house seemed to be bulging with presences coming up the stairs. She was terrified but managed to get to her feet to leave the room. She looked out the window and there she saw an unidentified flying object glowing with grey light even though the light did not illuminate the landscape. She panicked and does not know what happened after that until she was downstairs in her parents' bedroom. Over the years she has come to conceptualize her experiences as revealing to her dimensions of reality that she had not previously known to exist. Indeed, John Mack, who has studied them, has said that alien abduction experiences shatter experiencers' ideas about reality and engage them in a process of radical self-transformation thereby challenging a materialist world view (Mack, 1999).

Of course, in each of these three situations, whereas experiencers themselves are convinced, we can question the veracity of their convictions. We do not have direct access to these people's private experiences and are forced to rely on their reports. How reliable are these reports? And how can the sense of increased meaning or revelation of previously inaccessible dimensions of reality be validated? This is an epistemological dilemma. We can dismiss these types of observations as hallucinations or mental disturbances, as indeed materialists have often done, and ignore their ontological implications (Barušs, 2000). However, what if we suspend judgment for a moment and examine these experiences more closely with regard to their truth value?

Establishing Truth

Perhaps the first thing to note about what is true is that we have already discussed objective evidence to show that reality does not work the way that we ordinarily suppose. So who is to say what else is not possible? Our everyday convictions are wrong. Of course, by the same token, our extraordinary convictions could also be wrong. And indeed, often they are wrong. There really are mental disturbances in which people are mistaken about their ideas about verifiable aspects of reality. Are we doomed, then, to a form of epistemological relativism in which no one knows the truth? Perhaps, but if so, then at least there is some structure among the relative truths provided by dialectical reasoning.

Dialectical reasoning, as I am using that expression, refers to a way of thinking whereby a person proceeds from what she knows, through the addition of new knowledge, and the formation of a new synthesis (cf. Laing & Cooper, 1964/1971; Sartre, 1960). In effect, we move from lesser toward greater levels of complexity although this does not preclude the possibility of finding simple structures in some of the syntheses. In the case of this book, we started with materialism, added information about quantum weirdness and other anomalies, and came up with ... what? We have not come up with an alternative yet. But this raises another important point. We may need to suspend judgment for considerable periods of time while old paradigms about reality are being eroded and new world views are being assembled. The point is that knowledge, in principle, is cumulative, and in that sense nominally structured. I believe that we are closer to the truth whenever we have a more comprehensive understanding of a situation that better takes into account its disparate aspects.

I conceptualize dialectical reasoning as proceeding either horizontally or vertically. When it proceeds horizontally, we are learning more about the same kinds of things, whereas when it proceeds vertically, then we are moving toward an understanding of the principles underlying what we already know. Once a body of knowledge has been understood, it may become meaningless to review other horizontal versions of that same body of knowledge. For example, various researchers have come up with different breakdowns of the referents of the term consciousness, often without awareness of the efforts of others to do the same (Barušs, 1987). Once a researcher has in some manner conceptualized the differences between the various meanings, little is gained by reviewing others' efforts to do so. Or to take a more immediate example, once it is clear that materialism is an incorrect interpretation of reality, there is little to be gained by reading about other instances in which it fails. Indeed, after spending some ten years involved in the debate about the inadequacy of materialism, there was no point in my continuing to cycle through variations of the same arguments (Barušs, 1993; also Barušs, 1990b; 1992; 1995; 1998). I felt that my time could be better spent trying to understand what reality was really like. In other words, vertical advancement can be more meaningful than indefinite horizontal assimilation. This turns out to be an important ingredient in the application of science as a spiritual practice as discussed in Part III.

Knowledge also has the quality of being personal in that each person has her own understanding of reality which differs in insignificant or significant ways from that of others. A woman who has given birth to a child knows from her own experience something about what childbirth is like that no man can ever know. Or at least, not through any ordinary means. But childbirth also has a publicly observable aspect to it so that onlookers can have some sense of what is taking place. In the case of near-death, psychedelic, and alien abduction experiences, the publicly available view is so minimal as to be largely worthless for making judgments about the nature of such experiences. But for the person whose experiences they are, there has often been an apparent increase in her understanding of reality.

Raymond Moody, one of the first to systematically study near-death experiences, has used Plato's analogy of the cave to describe the effects on a person of re-entering this life after a near-death experience (Moody, 1988). In Plato's *Republic*, Socrates describes a situation in which men have been chained since childbirth in an underground cave so that they are facing the back of the cave and can not turn their heads to see their compatriots or the opening to the cave. Behind the men is a fire and between the men and the fire is a walkway with a low wall. Men are walking along the walkway, sometimes talking, carrying vessels or figures of various kinds that appear above the wall. Yet the men who are chained can see nothing but the shadows of each other and of the objects that are being carried behind them. And because there is an echo in the cave, they attribute the voices that they hear to the shadows that they see in front of them.

Now suppose, says Socrates, that the prisoners in the cave were to be released and turned around so that they could see the actual situation. Any of them would be perplexed and unable to name the objects being carried along the walkway. Furthermore, if he were to be dragged up a steep incline out of the cave into the sun, he would find that to be a painful experience. However, once acclimatised to the sun-lit real world, the former prisoner would no longer have any desire for the old life of the cave. If he were to find himself back among the prisoners, it would take time for his eyes to readjust to the dim light, and if, in the meantime, he were required to compete with the other prisoners in describing the shadows, he would be unable to do so. The others would conclude that it were better not to even think of going up out of the cave and would put to death anyone who tried to release another of the prisoners (Plato, 1968).

I think that the point of this analogy is that our ordinary interpretation of the world is seriously wrong. We are the walking dead busy deciphering the whisperings of the shadows. We are the mentally ill suffering from a mass psychosis. It is we who mistake what we experience through the senses to be the whole of reality. The realization that there is more to life is the epiphany that experiences in some alterations of consciousness can create for the person who has them. It is not that the old ways of knowing things have disappeared but that there has been an addition of knowledge causing the old views of the world to be revealed in a new light. Thus we have two pathways from materialist to transcendent beliefs: the one through an objective analysis of empirical data and the other through the compelling quality of personal experiences.

Reconstructing Science

Perhaps the reader may be thinking that it is one thing to consider objective evidence for the presence of anomalies but quite another to lend credibility to the florid speculations of those who imagine that they have been abducted by aliens. The whole point of science, after all, has been to place knowledge on more solid footing than people's whimsical musings and pathological delusions. But the personal and scientific ways of knowing are not as different as they might at first appear. Or, to be more precise, they converge on a single way of knowing when each is authentically realized.

The objective of science is to turn contentions about the nature of reality into questions and then to find ways of answering those questions. Hence, the point is to know something, rather than just to have ideas about it. A methodology is introduced in order to enable the researcher to get at the best answers possible. And the resultant data are used to inform the researcher's understanding of the nature of reality, thereby modifying her world view. That, to my mind, is how authentic science functions.

Sometimes this sequence becomes reversed or there are other distortions of the practice of science. The world view becomes predominant and impervious to change. The methods of science become rigidified, precluding the investigation of whole areas of inquiry that lie outside the scope of those methods. And the accumulation of uncontaminated information, rather than the pursuit of understanding, becomes the purpose of science. Under those conditions, science becomes scientism, a restrictive doctrine prohibiting authentic exploration (Barušs, 1996.).

As an example of the difference between authentic and inauthentic science, we can note that it is possible to have opinions about the survival of consciousness after death: consciousness does continue after death or consciousness does not continue after death. For authentic science that becomes a question: Does consciousness survive death? Having asked the question, we would need to find suitable methodologies for investigating it. Any data gathered from those investigations would be used to change the researchers' understanding of the nature of reality (e.g., Braude, 2003). But the world view of scientism is that of materialism and the belief that consciousness does not survive death does not get questioned. Ironically, the politics of science are such that those who persist in raising the question run the risk of being criticized for being unscientific.

For scientism, the accumulation of data requires adherence to rigid methodology presumably in order to ensure that the data remain uncontaminated with falsehoods. This need not pose a problem except in areas of inquiry where such methods can prohibit adequate investigation. One of the rules that ensures the integrity of the body of scientific information is the requirement that phenomena of interest be publicly observable. This rule has been part of the methodology of scientific psychology since the behaviourist revolution.[6] But what happens now if we are interested in a person's subjective experience? Well, we can look at the behavioural manifestations of such experience including a person's verbal behaviour. But can we do any better than that?

A person can make observations within the domain of her own private experience which can then be regarded as data for her. On the basis of these data, she can develop her own private theories (Mandler, 1985). Or public theories, for that matter. The philosopher Daniel Dennett, for example, has looked to his own experience to validate his cognitive theory of mind. His theory, he has said, is borne out by examining the situation in his own case (Dennett, 1978). Whether or not it is labelled as such, this is just introspection, the observation of one's own mind. But introspection, as an activity, has had a thorny history and it is not clear to what extent it has been explicitly accepted in science as a research strategy for understanding human experience. What is clear, however, is that there is considerable support for its use. In a survey of participants at a scientific conference about consciousness in 1996, Moore and I found that 93% of 212 respondents to our questionnaire thought that introspection was necessary for the investigation of consciousness (Barušs, 2003; Barušs & Moore, 1998). Whatever its official status, researchers, such as Dennett, have been using introspective data to validate their theories.

The reader may have noticed that, with the notion of introspection as a research strategy, we have come full circle back to the substance of the discussion about the compelling quality of personal experience. *Learning from one's own experience can be regarded as a scientific activity.* Hence, the path of personal conviction is compatible with an authentic scientific approach to knowledge. However, there are a few important points to note about regarding personal exploration as part of science.

The first point is that, given the presence of differences in investigators' experiences, scientific knowledge no longer constitutes a unified corpus, something that is sometimes considered to be a hallmark of science (cf. Bauer, 1992). Of course, there never has been such a unified corpus in science considering the amount of disagreement that exists. The controversies surrounding the anomalous phenomena with which this part of the book opened are just one example of such disagreement.

There is a second, more important point to make. The corpus of scientific information, in itself, is inherently meaningless. Information is certainly available in scientific journals and other data bases, but this information serves the individual scientists who seek knowledge. *It is their understanding that has meaning and not the scientific information as such.* The reader may have caught the movement toward a more transcendent conceptualization of science in this shift of focus from the meaningless to the meaningful. But whether or not one accepts such a shift, it does not change the fact that differences in what individual scientists think they know creates a stratified scientific community. And that, even if not explicitly acknowledged, in some cases data from scientists' own experiences appear to contribute to the stratification.

Where we end up, in effect, is with the notion that scientific knowledge, in its essence, is just dialectical reasoning, in that making sensory observations, compiling data, and theorizing are *ongoing noetic activities* for a researcher. Experiences that can be conceptualized as formal scientific investigation commingle with experiences that are not labelled as being scientific and both contribute to the development of a scientist's understanding. This dynamic is illustrated in the next paragraph in an account of the acquisition of my own understanding of anomalous resonance between portable REGs and entranced groups of people.

I had been familiar with the research concerning anomalous human–machine interactions at the Princeton Engineering Anomalies Research laboratory at Princeton University where it was found that REGs deviated from chance expectation in the direction intended by a human operator. I had read the technical reports of the research,[7] talked to those who had worked in the laboratory, and gone on a field trip with a group of my students to participate in the research. The findings had been intellectually convincing. However, when Nelson showed me the unambiguous deviation from chance of the behaviour of the portable REG that had occurred while a group of us had been watching a sunset, I was struck by the results. Such personal experience, while not in and of itself evidence of an anomalous occurrence, gave meaning to the more formal data. Both formally scientific and unscientific experiences have contributed to my understanding of reality. Hence, my scientific knowledge is subsumed as part of my knowledge more generally. And, of course, others who have not had the same experiences as I have had, will have different views about the nature of reality.

A third point to note is the need to make judgments about validity. In what sense is someone's learning from her experience really knowledge? What is to keep the dialectical process of increasing understanding from being a process of increasing delusion? These are difficult questions to answer. Simply living one's experience is not a guarantee of knowledge. Nor does reification of understanding in terms of information, as a materialist might tend to do, solve the epistemological problem, given that it is not clear that such reification is justified (cf. Fodor, 2000), and given that the meaning of information in such explanations is itself unclear (cf. the meaning in communication theory as given in Shannon & Weaver, 1964). Rather than trying to comprehensively resolve this problem, we can simply regard understanding as a primitive concept, whose referent can only be identified and validated by each person for herself through a process of self-examination.

There is a practical solution to the problem of the validity of self-knowledge, one to which I have already alluded, namely, that disciplined use of the mind is necessary in order to at least hope to separate knowledge from delusion. But what does that entail? I think that, at a minimum, disciplined use of the mind involves logical reasoning of some sort, at least up to a point. Furthermore, when gathering data about oneself, one can try to be systematic and keep as detailed documentation as possible. And how are we to regard the apparently noetic value of anomalous experiences? What do we do with apparent knowledge that is neither data acquired through observation nor a result of logical reasoning? Is it really knowledge? In our 1996 survey, Moore and I found that 69% of 212 respondents agreed that there are latent modes of understanding within a person that are superior to rational thought (Barušs & Moore,

1998). Thus, many of us certainly think that some kind of inner knowledge is possible. But let us leave this train of thought for the moment and take it up again in Part II. For now, let us just note that we have established the failure of materialism but not the failure of authentic science. And authentic science converges with personal exploration.

Self-Transformation

In discussing beliefs about consciousness and reality, we considered two aspects of conservatively transcendent beliefs found by Moore and myself: meaning and religiosity. I have said a little bit about meaning, but nothing yet about religiosity. Scientists generally appear to have little use for religion and may feel embarrassed to even read about it. The term "religion" itself, which had a broader meaning in the past, has come to refer more to institutionalized religious practice, whereas the word "spiritual," also associated with religiosity, has a more expansive meaning in terms of searching for a connection to something deeper and more meaningful in life than ordinary everyday existence (Pargament, 1999). There has been a decline in institutionalized religious practice in North America since the mid-twentieth century (Bibby, 1987; Wuthnow, 1998); nonetheless, according to one poll, 58% of Americans are seeking spiritual growth (Hood, Spilka, Hunsberger, & Gorsuch, 1996). What was previously a regulated activity has become a more liberalized and personal one (Taylor, 1999). For some scientists, however, the notion of spirituality may be no more palatable than that of religion. I had considered avoiding the words "religion" and "spirituality" altogether in this book so as not to alienate such scientists, but the use of these terms is well established in the academic literature. Thus, in particular, I use the phrase "spiritual practice" to denote any activity deliberately undertaken by a person for the purpose of seeking something more existentially meaningful than ordinary everyday existence.

One way to think of religiosity is to look at religious orientations. I am going to use an older characterization that has been modified in light of more recent research but which I find conceptually useful in its original form (Spilka, Hood, Hunsberger, & Gorsuch, 2003). Those who have an *extrinsic* orientation to their religious practice are interested in its utilitarian aspects and use religion as needed to serve their own ends. Those with an *intrinsic* religious orientation actually believe the doctrines of their religion and view religion as a framework for living their lives (Allport & Ross, 1967). There is a third orientation, *quest*, which is characterized by an open-ended and self-critical exploration of existential questions. Those with a quest orientation regard religious doubt as positive and resist reducing the complexity of life (Batson, Schoenrade, & Ventis, 1982/1993; Hood, Spilka, Hunsberger, & Gorsuch, 1996).

I was raised as an active participant in the Christian church. However, even as a child, I questioned the dogma that I was supposed to believe until, in my teenage years, I realized that I was trying to force myself into believing something that I did not believe. At that point I decided that it was more honest to give up trying to psychologically manipulate myself and that, rather than attempting to believe things, I would orient myself on the basis of my own understanding of reality. I could better conceptualize this shift a few years later when I studied Martin Heidegger's *Being and Time*, in which he has laid out distinctions between inauthentic and authentic ways of being in the world (Heidegger, 1926/1962). We are inauthentic when we get caught up in what we are expected to believe, and authentic when we base our actions on our own understanding (Baruš, 1996). I went from an unsuccessful intrinsic orientation to a quest orientation and from inauthenticity to authenticity. The interest that I had had in existential issues had not changed, it was just that the manner of dealing with them had become one of open inquiry.

The shift from an intrinsic orientation to quest applies also to science. There are true believers of scientism, those who honestly believe that the world is a material place. Having once gone through the process of divesting myself of the restrictions of a religious doctrine, I have no interest in trying to envelope myself in the dogma of materialism. But materialism is only necessary for inauthentic science. Authentic science is an open inquiry, including an inquiry into the degree to which materialism is a correct interpretation of reality. There is a parallel here between science and religion in that both admit of having believers as well as those who are on a voyage of discovery. In that sense, the dynamics of scientific exploration and quest converge whenever science reaches into the existential terrain.

Ordinarily we are not concerned with questions about the deeper significance of life. But what if we were? Perhaps questions arise as a result of personal crises that break our preoccupation with everyday matters. What is the meaning of life? What is the purpose of our existence? These questions are rationally coherent, but there are no readily available rational answers. In fact, it may not be possible to find answers rationally. That is not to say that we are necessarily incapable of resolving such questions. But what may be required could be a process of self-transformation whereby latent psychological faculties become activated with which the answers can be known. Or to put it more colloquially, one needs to become enlightened — to experience a transcendent state of consciousness in which normally inaccessible knowledge becomes available. Whether or not such self-development is possible is an open question, but the answer to that question lies in exploring the notion of spiritual growth, and not in making pre-judgments about it.

Let me make some additional comments about the practical significance of spirituality. It is not just that spirituality becomes relevant in the face of existential crises, but there has been accumulating evidence that religiosity is associated with a range of potential benefits for a person. One of those benefits is greater physical health. For example, religiousness is associated with low rates of cardiovascular disorders and cancers although the psycho-physiological mechanism through which these effects occur is not known. There has been speculation that those who are more religious have better health habits, in part because of prohibitions against destructive behaviours such as drug addiction, and proscriptions of health-promoting behaviours such as seat-belt use (Hood, Spilka, Hunsberger, & Gorsuch, 1996). It is also possible that the association between religiosity and health could be mediated by the presence of positive emotions such as gratitude. In that regard, it should be noted that the presence of gratitude in one's daily mood is correlated with an intrinsic religious orientation but not extrinsic or quest orientations, although those who experience spiritual transcendence are also more grateful (Emmons & Paloutzian, 2003). Moreover, considerable empirical research has shown that the practice of some types of meditation can induce a relaxation response which may, in part, contribute to improvements of various medical conditions for those who practice such meditation techniques. Because of its potential beneficial effects, meditation has been recommended by some health professionals as a therapeutic strategy (Andresen, 2000). Thus there are diverse ways in which religiosity could be beneficial to one's health.

I should qualify the link between religion and well-being by pointing out that religiosity can also be harmful. For example, the same qualities of control and self-esteem that are associated with religiosity can contribute to harmful behaviour such as aggression. And there certainly appears to have been considerable religiously motivated aggression, for instance, in the form of holy wars (Bowker, 1997). There are also associations between religiosity and psychopathology which we will consider in the next section. What these qualifications suggest is that there may be more beneficial and less beneficial ways of being religious. In particular, religious fundamentalism, rigid belief in a single inerrant version of reality, has been characterized in pejorative terms as lacking cognitive complexity and having a "low level of spiritual maturity" (Spilka, Hood, Hunsberger, and Gorsuch, 2003, p. 390).

We noted earlier that intelligence, as it has been conventionally conceptualized, has been associated with more transcendent beliefs. However, intelligence has also been regarded as having multiple forms, so that, for example, a person could have musical intelligence or interpersonal intelligence (Gardner, 1983; 2000). If some types of religiosity are adaptive behaviours associated with success in various areas of life, including greater physical well-being, then why not regard religiosity as a form of intelligence? And, indeed, the notion of spiritual intelligence, sometimes designated as "SQ", has emerged as a psychological construct (Emmons, 1999; Zohar & Marshall, 2000), although the question remains whether SQ is a truly separate intelligence or just a domain of activity that engages more conventional forms of intelligence (Gardner, 2000).

Robert Emmons has identified five characteristics of spiritually intelligent individuals. The first is an ability to transcend the limits of ordinary physical reality. The second, which is related to the first, is the ability to experience transcendent or mystical states of consciousness. The third is the ability to infuse everyday life with a sense of the sacred. The fourth is the ability to harness spiritual resources for problem-solving. And the fifth is the ability to be virtuous (Emmons, 1999). Cast in the form of such an intelligence, spirituality is a desirable aspect of oneself.

With regard to the first and second of these characteristics, we have considered some examples of experiences that can lead to transcendent beliefs, some of which can be identified as transcendent experiences themselves, such as near-death experiences (Ring, 1987). But the prototypical transcendent experiences are mystical experiences. These are experiences in which a person feels a sense of unity with all of reality, transcendence of the boundaries of time and space, and deeply felt positive emotions such as love and peace. There may also be a sense of sacredness; a feeling of paradoxicality, in that aspects of the experiences are felt to be true in spite of violating ordinary logical principles; there may be subsequent positive changes in attitudes or behaviour; and such experiences may be transient and ineffable. Perhaps most notable for the purposes of our discussion is that mystical experiences have noetic value. In other words, those who have them typically believe that mystical experiences convey knowledge about the nature of reality (Barušs,

1996; 2003). In that sense, spiritual intelligence is also about cognition. We shall take up the notion of seeking transcendence in Part III.

Spirituality and Psychopathology

It should be noted that religious and spiritual experiences have not always been regarded as positive events and so it would be prudent for us to consider in greater detail the relationship between spirituality and psychopathology. We have already seen that belief in the occurrence of some anomalous phenomena has been considered to be a symptom of a mental disorder. From an early psychoanalytic point of view, religious experiences are the result of misattributions. In particular, regressive experiences of unity in an infantile state of mind get erroneously interpreted as union with God. And because a person has improperly identified the actual nature of her experiences, she can be regarded as being delusional and hence as suffering from a psychopathology (Hood, Spilka, Hunsberger, & Gorsuch, 1996).

There are also several theories as to how mystical experiences could arise as irregularities of brain function. For example, according to Arnold Mandell, the activity of the neurotransmitter serotonin is disturbed by meditation, stress, or drugs such as psychedelics in such a way as to fail to inhibit the firing of CA_3 cells in the hippocampus of the brain. The resultant hyperactivity of this layer of cells prohibits the proper comparison of incoming information with that which is stored in memory resulting in a sense of unity with everything. The hyperactivity leads to "hippocampal-septal synchronous discharges" which get interpreted as positive emotions such as ecstasy. Some of the affected brain cells eventually die, creating personality changes associated with religiosity such as good-naturedness, emotional deepening, and the occurrence of "transcendent consciousness". These experiences of pathological brain events get erroneously attributed to God (Mandell, 1980; quotations from p. 400). Again, a person is believed to be deluded if she mistakenly misattributes aberrant biological activity as meaningful enlightenment.

However, efforts to find correlations of psychopathology with the occurrence of mystical experiences have been unsuccessful, although, perhaps not unexpectedly, correlations of psychopathology with diabolical experiences have been found (Caird, 1987; Spanos & Moretti, 1988.). The finding of neurophysiological mechanisms corresponding to mystical experiences would also be expected, although there are various quite different versions of what those mechanisms could be (eg., Mandell, 1980; Persinger, 1987; d'Aquili & Newberg, 1999; 2000). The presence of neural correlates does not invalidate mystical experiences any more than finding the neural correlates of other types of experiences invalidates them (Barušs, 1996). Furthermore, as we have noted previously, there is considerable evidence for the beneficial effects of religiosity and spiritual experiences, so that it is not logical to regard them as being necessarily dysfunctional.

But the interface of religiosity and spirituality with psychopathology is more complex than that. There can be psychological disturbances precipitated by transcendent experiences, there can be mixed states with both exceptional and pathological elements, and apparently genuine spiritual insights can emerge in pathological conditions. Let us briefly consider each of these in turn.

Whereas religiosity and spirituality are generally associated with psychological well-being, transcendent experiences can have disruptive effects on people's lives. This has been noted, for example, with near-death experiences, which can result in fear of ridicule, anger, depression, alienation from relatives, broken relationships, and career interruptions. It appears that these problems can be attributed, at least in part, to an experiencer's more transcendent beliefs clashing with the more materialist beliefs of those around her (Greyson, 2000). This is the problem of re-entering ordinary life after having had a glimpse of other possibilities as illustrated in the analogy of Plato's cave. More dramatic disruptions can result from alien abduction experiences, which, as noted previously, can have a profound impact on experiencers' ideas about reality. Alien abduction experiences have been associated with intense fear. In one study, when asked to rate the fear that they had experienced at the time of abduction from 0 to 10, experiencers actually rated their fear from 10 to 100 (McLeod, Corbisier, & Mack, 1996). Although experiencers do not have any symptoms of known psychopathologies that could account for their experiences, they do appear to be suffering from the after-effects of trauma (Appelle, Lynn, & Newman, 2000; Mack, 1994). More generally, spiritual self-transformation has sometimes been regarded as a pleasant activity without adequate appreciation for the disruptions that psychological restructuring can have on a person. I have frequently seen disruptions such as confusion, depression, and difficulties with social adjustment, among those who awaken to their latent possibilities and undertake a process of self-transformation (Barušs, 1996).

In some sense, a psychotic is a failed mystic (cf. Assagioli, 1965). Perhaps the same latent capabilities are activated in both psychosis and mysticism but, in the case of psychosis, a person is psychologically unable to integrate the experiences. David Lukoff was a doctoral student in anthropology at Harvard University when he dropped out, hitchhiked across the country, took LSD, and wanted to start a growth centre. At one point he noticed that his hand was glowing and became convinced that he was the reincarnation of the Buddha and the Christ. He thought that he was becoming spiritually enlightened and wrote a book that he believed would lead to planetary spiritual renewal. After some six months of this alternative cerebration, he reconceptualized his actions as folly and reoriented himself to a more conventional lifestyle. Subsequently, Lukoff became a psychologist who was instrumental in bringing attention to mystical experiences with psychotic features and in embedding religious or spiritual problems as a non-pathological condition in the *Diagnostic and Statistical Manual of Mental Disorders* (American Psychiatric Association, 2000). For Lukoff, his psychotic episode was not just an aberration to be forgotten, but an experience that could be mined for any meaningful insights that it might have to offer (Shorto, 1999; Lukoff, 1985; Lukoff & Everest, 1985).

Is it possible that spiritual revelations could occur even as a result of psychologically dysfunctional situations that are not usually conceptualized as being spiritual? The psychological construct of transliminality refers to openness to cognitive and emotional material arising from within or outside the mind (Thalbourne, 1998). A disruption of the mind could increase *transliminality* giving a person access to previously unavailable resources. That is to say, a breakdown of the psyche could create conditions that activate aspects of it that would otherwise not be revealed. For instance, given the intense fear experienced by those who have come to believe that they have been abducted by aliens, is their subsequent spiritual transformation a result of the fear? In fact, is fear the only factor or is there something about the nature of alien abduction experiences themselves that contributes to the occurrence of constructive psychological changes (Mack, 1999)? In dissociative identity disorder a fragmentation of self-identity occurs so that a person may appear to have different personalities in different situations. Some of these personalities have been identified as demons or a person's dead relatives (Ross, Norton, & Wozney, 1989). The somewhat disturbing question arises of whether these alternate personalities really are manifestations of demons and dead people. Has the trauma that is usually associated with the etiology of dissociative identity disorder opened up a person's psyche, making it porous to ordinarily unseen influences outside of itself (Barušs, 2003)? Such notions of possession may seem far-fetched but they present us with some speculative possibilities to which we will return again in Part II. At any rate, once a person has been labelled with a mental disorder, any genuinely anomalous phenomena that may occur for her would likely remain unrecognized by others who would be unlikely to distinguish it from the manifestations of her illness. The point here is that psychological disturbances might unearth aspects of the psyche that could contribute to spiritual growth. As we can see, there are complex relationships between transcendent and pathological experiences. But transcendent experiences are not themselves dysfunctional states of mind.

A Transcendent Interpretation of Reality

If materialism is false, then what is true? I am not certain how well that question can be collectively answered at this point in the history of our culture. Some people with transcendent beliefs think that they have the answer, but there have been no theories with universal explanatory appeal. The emphasis seems to have been on trying to deny the existence of anomalous phenomena or making various materialist interpretations work. Let me propose at least the beginning of a line of development for a possible theory by returning to some ideas from quantum mechanics and exploring their relevance to consciousness.

A state vector in quantum mechanics is a mathematical description of the properties of interest of a physical system of fundamental particles. The problem is that the state vector cannot predict precisely what will be found when the physical system is actually observed. What the state vector does, in effect, is to give us the *probabilities* of a number of possible impressions that could be created on an observer. Upon observation, the state vector collapses so that those possibilities become a single physical event.

There are various ways of thinking about bridging the state vector description of matter with the existence of actual physical objects. According to the minimal interpretation of the quantum mechanical formalism, the state vector is just a device for making calculations about experimental results. Since such experiments involve the observation of macroscopic devices, such as cloud chambers in which subatomic particles can leave visible tracks, we never really have to depart from the domain of people-sized events to make statements about subatomic physical objects (Sudbery, 1986; Wigner, 1961/1983a). Such a resolution is incomplete given that the fundamental particles are supposed to be the constituent parts of people-sized things and hence should form the basis for an explanation of the experimental processes themselves, which otherwise remain inadequately explained. Another way of bridging the gap has been to posit that environmental effects wash out all but one of the possibilities inherent in the state vector through a process known as "decoherence," although it does not appear that they can always thus be eliminated (Polkinghorne, 2002; Adler, 2003; quotation from Polkinghorne, 2002, p. 44).

Let us consider yet another way of collapsing the state vector. Let us supplement the state vector description by postulating the existence of hidden variables that determine what actually occurs during an observation. We saw that hidden variables could not be used in a simplistically deterministic way for describing the behaviour of particles since Bell's Theorem appears to be violated. However, maybe the hidden variables do not reside in ordinary objective reality but in the interface with a transcendent domain. Maybe the hidden variables are *volitional directives* emerging from a transcendent aspect of reality that determine the actual manifestation of matter upon observation. Clearly, it seems reasonable to suppose that any such directive agency would operate not only during formal experiments in physics, but in an ongoing manner in our everyday lives to produce the reality that we actually experience moment by moment. The tendency has been to say that it is consciousness that is such a hidden variable — that the act of observation by a conscious agent collapses the possibilities given by the state vector into the reality that manifests itself. Perhaps that is true in some cases (Schwartz, Stapp, & Beauregard, 2005). In general, however, we do not go about our lives deliberately intending particular events to occur, so the volitional directive cannot always just arise from consciousness in its ordinary manifestation as a stream of sensations, feelings, and thoughts that occur for us. Rather, if there is an association with our consciousness, it appears that it may be through some deeper aspect of it to which we do not ordinarily have ready access.[8]

Some version of such a hidden-variables theory can only be true if there is some domain of reality that is transcendent relative to ordinary objective reality, yet can have causal effects on it. Transcendent beliefs, as Moore and I have found in our surveys, include the notion that there is a universal consciousness that gives rise to physical reality. This corresponds, on the face of it, to an idea proposed by physicist David Bohm that there is an implicit order to the universe underlying its explications as both physical manifestation and consciousness (Bohm, 1980/1983). Such a theory could account for the synchronicity between some mental and physical occurrences such as the deviations from statistically expected behaviour of the REG during the Lake Huron sunset. While the source of the volitional impetus in the collapse of the state vector, the universal consciousness of transcendent beliefs, and Bohm's implicate order may actually be separable aspects of reality, let us use the expression "deep consciousness" or "transcendent consciousness" to refer at least to the direction in which these concepts are pointing.

Positing the presence of deep consciousness implies the existence of aspects of reality to which we do not have direct sensory access. I have already argued that reality is not exhausted by that which is sensorially apparent. The analogy of the light spectrum is sometimes used to illustrate our inability to perceive all of reality, in that we can see only a small fraction of light frequencies, while most of them remain unseen. Despite the fact that we cannot see all the light that is present, it can nonetheless physically affect us, for example, in the form of ultraviolet rays or x-rays. More generally, we could experience various influences that lie outside the receptive range of our sensory systems. The question is, what access can we have to such deeper aspects of reality?

There is a theory of personality known as psychosynthesis, developed by the Italian psychiatrist Roberto Assagioli, which bears on the question of access to deep consciousness. In psychosynthesis, the psyche has various components. One such component is the awareness of our ongoing experience. At the centre of this awareness is the conscious self, the one for whom the contents of consciousness occur. There is a preconscious, consisting of material that is readily available to our awareness should we direct our attention to it. There is a subconscious, not readily accessible to our ordinary awareness, whose composition includes biological drives, results of past learning, buried memories, and psychopathological material. And, what is particularly relevant to our discussion, there is a superconscious, consisting of a part of ourselves that is wiser and more emotionally mature than we ordinarily are, from which intuition, moral imperatives, genius, and spiritual insights can originate. As with the subconscious, the superconscious is not immediately accessible to our ongoing stream of consciousness, although it can sometimes manifest itself, such as during mystical experiences. At the apex of the superconscious is the higher self, which is our self, in the sense that the higher self can potentially form the basis of our self-identification. The entire psyche is porous so that influences from outside a person can impact on a person's psyche without the mediation of the senses (Assagioli, 1965; 1973/1974; 1988/1991; Baruss, 1996; Ferrucci, 1982; Firman & Gila, 2002; Hardy, 1987).

In psychosynthesis, the self is not an autonomous part of one's psyche, but a reflection of the higher self, so that there is an inherent link between the level of reality of ordinary consciousness and that of mystical experiences. But now we have a means of access to deep consciousness to the extent that the consciousness of the higher self is at least in the same ontological vicinity as deep consciousness. In other words, the idea here is that there are aspects of reality, including transcendent aspects of consciousness, from which volitional directives can emerge to structure physical reality, and to which we individually might have access within our own psyches. This is all speculative, I know, but at least we have a somewhat minimal interpretation of reality that we can use in our further discussions.

We started out by noting that materialism is dead. This is the case because matter does not conform to a materialist interpretation of reality as revealed by violations of Bell's Theorem and the wave-like behaviour of subatomic particles. There is also some objective evidence for the occurrence of presentience and anomalous resonance, both of which strain materialist and physicalist theories. The persistence of materialism can likely be attributed to the politics of science whereby open examination of the prevailing doctrine of materialism is tacitly prohibited. Beliefs about the nature of reality can range along a material-transcendent dimension and materialist beliefs, on the part of scientists, may contribute to the persistence of materialism. However, among undergraduate students it has been found that a more rational approach to the world and greater intelligence are correlated with transcendent rather than materialist beliefs. Indeed, given the benefits associated with at least some forms of religiosity, the ability to be religious has been conceptualized as a type of intelligence —spiritual intelligence. But what of scientists who are not interested in conventional religion? There is a religious orientation, that of quest, which essentially coincides with authentic scientific exploration. Such exploration is a dialectical integration of formally scientific and personal experiences for which a disciplined mind is necessary. Authentic science can be concerned with existential issues but in that case the research consists of self-exploration and self-transformation. And the goal? Some form of transcendent consciousness, perhaps, whatever that may turn out to be. But before seeking transcendence, I want to return to an examination of the nature of knowledge, and, in particular, of inner knowledge.

Part II

Access to Inner Knowledge

In a journal entry for October 10, 1976, I have written three equations describing teleportation. "What?" you ask. "Teleportation?" Let me explain how they came about.

Five of us who were spiritually inclined would get together once in a while to meditate. One evening we lay down on our backs on the floor of a room with our heads together so that our bodies formed a five-pointed star. We closed our eyes and deliberately relaxed. Then one of us spoke an induction lasting about 10 minutes in which we envisaged ourselves going into a heightened state of awareness. After a period of silence we described to each other what was happening to us. On this particular occasion, we started to see images pertaining to teleportation. Or at least, four of us were getting information about teleportation, while the fifth remained silent. As the session progressed, we pieced together the following story.

There is an unmanifest world of which the physical world is a mirror reflection. What happens is that we continuously flicker back and forth between the unmanifest and the physical worlds. The flickering is so quick that we ordinarily do not notice it. We surmised that this same process occurs with an electron. There is a probability cloud that determines where an electron can appear at any given point in time. It appears. But then, it disappears again until some future time at which it appears in some location determined by the probability cloud, but without any flight path from the previous location. An electron has little volitional control over its actions. Hence, it cannot manage to make it back to the same location each time. We, however, as human beings, have an expectation that we will end up where we had been previously, and so, through the inadvertent exercise of our will, we do end up in the same place from which we left, thereby creating continuity of spatial presence. However, we do not need to come back to the same place. While we are in the unmanifest domain, we can flip a switch, as it were, and come back anywhere at any time that we choose. For example, we could theoretically show up in New York and then, a while later, show up again back in Toronto. That is the manner in which teleportation would be possible.

There were several symbols for the relationship between the unmanifest and manifest worlds. One of them was that of a court jester with one leg black, the other leg white, and with lateral reversal of the black and white colours for the arms. The court jester holds his hands above his head with palms upturned. Unseen, palms against palms, is an upside-down court jester with the black and white colours reversed for the arms and legs. The idea is that we keep flipping back and forth between the right side up and upside down versions of the court jester without being aware that we are doing so.

We also "got" the three equations describing teleportation that I mentioned:

$$E = \sqrt{\frac{(scm)^2 + 1}{-(scm)^2 - 1}} \qquad V = C \circ \frac{(sm)^2 + 1}{-(sm)^2 - 1} \qquad C = V \circ \frac{(sm)^2 + 1}{-(sm)^2 - 1}$$

where C is the speed of light or consciousness, and sm is small mass. The symbols E and V presumably stand for energy and velocity, although the energy and velocity of what, is not clear. And I do not know what that c is doing in the middle of the sm in the first expression. Standard algebraic manipulations, if they were to apply, would reduce these equations to:

$$E = i \quad V = -C \quad C = -V$$

where *i* is the imaginary number that is the square root of -1. Such is teleportation as it was "revealed" to a group of us in altered states of consciousness.

Inner Knowledge

The example that I have just given is an example of what appears to be inner knowledge and, more specifically, of channelling, which we will define separately and consider in more detail in a bit. Inner knowledge, as I am using that expression, refers to knowledge that is not evidently the result of physical sensory perception or ordinary reasoning. It subsumes such putative processes as intuition, psychism, guidance, mystical revelation, enlightenment, and, as mentioned, channelling. These are not all the same, but they share the characteristic of referring to knowledge that occurs within a person through apparently extraordinary means (cf. Yewchuk, 1999; Harman & Rheingold, 1984). In Part I we considered the possibility that there could exist modes of understanding that are superior to rational thought; a superconscious that is wiser than we ordinarily are; direct introjection of nonsensory information into a porous psyche; and noetic enlightenment associated with transcendent states of consciousness. Does any such inner knowledge exist? And if it exists, how can it be accessed?

I think it is clear that if inner knowledge exists, and if such knowledge can provide insight into the nature of reality, then it would be beneficial to have access to it. Furthermore, it does not take long to realize that various spiritual teachings point toward the primacy of inner knowledge (cf. Ferrucci, 1982; Grabhorn, 2000; Roberts, 1991; 1993; Bailey, 1950; Merrell-Wolff, 1994; 1995b), so that inner knowledge, if it exists, supersedes book-learning.

But the proposal that inner knowledge exists is a controversial one. In my experience, mainstream psychologists generally consider that claim, or at least its more interesting variations, to be preposterous, whereas spiritual aspirants regard it as a fundamental fact. I do not think that looking within for answers is nearly as facile as it has been portrayed to be by some new age enthusiasts, nor is it altogether impossible either as conventional scientists would maintain. The subject area is a quagmire with little empirical research to guide us. But after having seriously explored it myself for 35 years, both academically and experientially, I hope that I can address some of the issues that can arise when trying to access inner knowledge and point the way toward further experimentation.

Rather than trying to give an overview and critical appraisal of the entire subject matter, I will take a particular course through this material to give some idea of the practical issues that one can face. In doing so, let us start by looking at some examples of intuitive functioning, raise the question of its validity, and examine contemporary explanations for it. We will follow that by considering channelling and some research to support the veridical nature of at least some channelled material. Then we will look at the phenomenology of intuitive functioning in the ordinary waking state as well as in altered states of consciousness, with particular attention paid to the role played by figurative symbols. I will describe some instances of inner knowledge in dreams, including the incubation of dream solutions to one's problems. Finally, I will try to draw some conclusions about access to inner knowledge. Throughout this part I will use examples from my own experience as we consider this material.

Knowledge versus Coincidence

Let us begin by considering a series of typical examples of phenomena that can be identified as instances of inner knowledge and then some of the problems that can be encountered in trying to ascertain their validity.

The first example is a somewhat prototypical incident taken from Lynn Grabhorn's book *Excuse Me, Your Life is Waiting*. Grabhorn has written that her car used a brand of synthetic oil that was difficult to obtain at the time of the event she was describing. She was taking a trip on a highway when she realized that, not only had she forgotten to add the much needed oil before leaving, but she had forgotten to bring any of it with her. She wondered what to do when she "got a hit" to take the next exit. This she did and came upon an apparently deserted town. She stopped, got out of the car, and noticed a dilapidated auto shop. It happened that the proprietors carried the exact type of oil that she needed (Grabhorn, 2000; quotation from p. 166). Interesting. However, it is difficult to glean much insight from this example of following a hunch. What choice did Grabhorn have but to pull off the road? And how unusual was it to find an auto shop that carried the right motor oil? Was this knowledge or coincidence? Or knowledge of coincidence?

During 1976 and 1977 I attended group meditation sessions with Alma. Alma was an outspoken woman in late adulthood who claimed to have been psychic since the time that she was an infant. And, while there were some unusual phenomena that occurred from time to time, I do not recall having any incontrovertible demonstration of exceptional abilities on her part. On the contrary, she made a prediction that the world would end in the early 1990s with the survival of only a small handful of those of us who were worthy. However, the world did not end in the early 1990s.

At one point I was staying in England by myself and feeling rather lonely. One Sunday afternoon I deliberated what to do. I had intended to go to a concert being played by a string quartet at a church several blocks from where I was staying. Nevertheless, as I was feeling out of sorts, I decided not to go. As I recall, I had a strong impression that I should go to the concert. However, I mistrusted the reliability of such intuitive impressions. Since I could see no particular value in going to the concert I decided to override the intuition and stayed in my flat. Several months later, when I was back in Canada, I was chatting with a friend who, it turned out, had attended that concert. What a coincidence! It would certainly have lifted my spirits to have spent some time with him.

I play ice hockey year round, usually several times a week. As I was driving to play one night I had a strong impression that I would be seriously injured during the upcoming hockey game. I try to be careful when I play, but with the presence of swinging sticks, sharp blades, colliding bodies, and a flying rubber disk, the possibility of injury is not just hypothetical. And indeed, over the years that I have been playing, I have witnessed many injuries to others. I explicitly identified my impression as an intuition and analysed it as I was driving. Given my continued mistrust of the reliability of such impressions, I decided to ignore it. Nothing untoward happened. Nothing at all.

All of these hunches and impressions have to do with future events, some of which did occur as anticipated and others of which did not. What I have found is that advocates of intuition have tended only to give examples of intuitive impressions that turn out to be correct and have failed to adequately portray the difficulties of finding one's way through an inner world that also contains instances of intuitive impressions that turn out to be wrong. I have tried to give a more representative selection here of what I think actually occurs.

When intuitions of this sort do turn out to be correct, how are we to know that they are not just coincidences that occurred by chance? However rare, statistically improbable events do occur. While it is important to ask this question, we need to be careful not to discount the possible occurrence of inner knowledge just because we realize that infrequent events occasionally occur. Accepting that a rare event has occurred does not mean that inner knowledge has not also taken place. There could be a small probability that one of my friends would show up in a neighbourhood in which I am living overseas and a small probability that I would have a spontaneous hunch to go to an event at which he was also present. The question then becomes, of whether there is some way in which that coincidence can be deliberately identified so that, in effect, inner knowledge has taken place. In other words, is it possible to deliberately take advantage of low probability events that occur by chance?

There are other complications raised by predictions of future events. For instance, it should be noted that, from a materialist point of view, there is no problem, in principle, with predictions of future events because all events in the universe are completely determined. The future cannot be other than it is going to be and so there is no theoretical reason why images of the actual future cannot occur in the present as part of a deterministic chain of events. In fact, such images, if they were to occur, would necessarily have to occur when they do. Incidentally, so would any impressions of choice that anyone might have. That is to say, from a materialist perspective, any impressions of future events or choices that we might make are themselves just necessary parts of a deterministic chain of events.[9]

Do we have real choice? Can predictions of the future serve to avert the occurrence of the predicted events? Within the sphere of human activity, it may be possible to effectively change one's course of action on the basis of intimations about the future. I doubt that that was the case with Alma's predicted end of the world or my intuitions about the threat of hockey injuries. I think that those were simply false signals. Or, more accurately perhaps, those were warnings about real possibilities that were coloured by too much conviction. However, should we be able to actually change future events that we can foresee, it would be difficult to prove that we had actually done so given that we would need evidence to demonstrate that our prediction had not been wrong in the first place. Finding our way through the vagaries of inner knowledge is not nearly as simple to do as it may at first appear.

Implicit Cognition

Let us look at inner knowledge for a moment from the perspective of contemporary cognitive science. In Part I we considered the possibility that we could distinguish knowledge from delusion through the disciplined use of the mind. At the core of such disciplined mental activity would be some sort of logical reasoning. However, the actual flux of our ongoing thoughts is neither particularly disciplined nor logical. Rather, we tend to automatically rely on reasoning shortcuts that often involve the use of simplified schemata for evaluating complex situations. In the context of making judgments, these reasoning strategies have been called "judgmental heuristics" (e.g., Cialdini, 1985/1988, p. 7), and, more generally, nonconsciously derived knowledge can be referred to as "implicit cognition" (cf. Dorfman, Shames, & Kihlstrom, 1996). I am using the term "nonconscious" rather than "unconscious" to distinguish the construct of the nonconscious from the meaning of the unconscious in psychoanalytic theory and intend it to mean approximately anything outside of our explicit awareness. Whatever rigorously logical structure our thinking has is a synthetic product of conscious and nonconscious processes.

From the viewpoint of cognitive science, all knowledge is based on sensory perception, memory, and implicit and explicit reasoning. There are no other sources of knowledge. In particular, all inner knowledge is the result of memory and ordinary nonconscious processes, regardless of how the knowledge appears to have been derived for the one for whom it occurs. Thus, hunches, intuitions, and other forms of inner knowledge are considered to be just the results of implicit cognition. And, as evidenced by empirical studies, such knowledge has been shown to be both strikingly accurate in some situations and terribly wrong in others (Myers, 2002). The idea, of course, would be to learn how to improve such intuitive functioning so as to be correct more often (Hogarth, 2001). Let us consider some of the types of examples that have been cited to illustrate inner knowledge as ordinary implicit cognition.

Paul Lewicki conducted experiments in which participants had to find a target character, such as the number 6, in a matrix of distractor characters on a computer screen as the target character changed location from one display to another. When the background pattern of distractor characters varied consistently with the location of the target character, participants were able to learn to find the target character more quickly than when there was no consistent variation of the location with the background. The learning, however, was not conscious. Despite hours of effort in some cases and the lure of a monetary reward, none of the participants could discern what the consistently varying pattern of the distractor characters was. This included Lewicki's professorial colleagues, whose performance improved with practice and deteriorated when the variation of the background pattern with the location of the target character was changed. This is a prototypical example of implicit cognition. We know something, but we do not know what it is that we know, even though it clearly affects our behaviour (Lewicki, Hill, & Czyzewska, 1992).

In the 1930's, egg producers in the United States wanted to raise only hens, so poultry owners needed to know right away when a chicken was born whether it was male or female. The problem was that the male and female organs of the chicks looked identical and it took 5 to 6 weeks for adult feathers to develop so that the sexes could be distinguished. However, it turned out that some Japanese had the ability to sex chicks that were one day old. Training programs were set up so that the Americans could learn from the Japanese. And the Americans did learn to distinguish the chicks at rates of 400 to 1,400 an hour by a quick look at the small differences between the genital eminences of the two sexes. But not all chicken sexors used the same technique, some of whom could apparently distinguish chicks accurately just by touch. This ability to sex chicks has been cited as an example of intuitive functioning (Lunn, 1948; Dreyfus & Dreyfus, 1986; Myers, 2002).

Another example is that of individuals with savant syndrome. These are people who excel at some tasks amidst general intellectual deficits usually in the context of autism (Heaton & Wallace, 2004). For instance, some savants can quickly name the day of the week for any date within a given range of years. In one case, a calendrical savant could name the correct days of the week for years up to 2,051,279.[10] Some savants can quickly compute square roots of numbers, make detailed drawings from memory, attain musical proficiency, and show exceptional artistic ability (Myers, 2002; Yewchuk, 1999). Typically, savants cannot articulate how they accomplish their feats. As one calendrical savant has said, "use me brain" (Yewchuk, 1999, p. 64). Whatever the exact cognitive processes may be that are involved in this syndrome, they appear to function implicitly without becoming accessible to awareness (Heaton & Wallace, 2004).

But implicit cognition can also get things wrong. Calendrical savants, who appear to use a combination of memorization and calculation, often get the days wrong for dates prior to 1752 when Great Britain dropped eleven days by following September 2 with September 14 in a transition from the Julian to the Gregorian calendar (Cowan, O'Connor, & Samella, 2003).

There are also numerous biases that systematically operate in our thinking. For example, there appears to be a truth monitoring function, which is continuously trying to discriminate incoming information in terms of its truth value. As you, the reader, read this book, you are making judgments about what is true and what is false about what I say. Your truth monitor could be set at an uncritical level so that you are willing to accept whatever I say as long as nothing raises your suspicion. Or it could be set at a highly critical level whereby every contention is examined for its weaknesses. Or it could be somewhere in between. However, just considering a contention can strengthen the subjective feeling that that contention is true (Fiedler, 2000). It is interesting to note in that regard that academics had to justify "references to the transcendent" when writing about religious activities in a scientific context (Dawson, 1987, p. 227). Was the justification necessary to counteract fears by materialists that the mere mention of a transcendent reality would make the scientific community susceptible to seriously considering the possibility of its existence?

Once beliefs have been formed, implicit cognition can keep them in place. For example, *confirmation bias* works by seeking out information that supports our version of reality and discounting contrary evidence. In one study, undergraduate university students were assessed regarding their belief in extrasensory perception. Then they were asked to evaluate either a favourable or an unfavourable report about extrasensory perception. Not surprisingly, those who received a report that was contrary to their own beliefs rated the report as being of poorer quality than those who received a report that matched their beliefs. There were no differences in actual critical thinking between those who believed in extrasensory perception and those who did not (Roe, 1999). Clearly, such biases also have implications for scientific activity.

To what extent is inner knowledge always just the result of ordinary implicit cognition? For instance, was my hunch to go to the concert by the string quartet really that surprising? After all, I had intended to go to it in the first place and getting out of the flat and listening to some music would have improved my mood. Perhaps a part of my mind had kept track of that and brought it into awareness. My friend's attendance could have been entirely coincidental. In the end, I do not know if there was anything extraordinary about that particular hunch. But I do think that we can account for many instances of intuitions, such as those given in the first section, as the effects of ordinary nonconscious processing.

Ironically, I am not convinced that all of the examples given as instances of implicit cognition are so easily explained. Although some inroads have been made in that regard, little is known about the actual cognitive mechanisms through which implicit cognition works (cf. Dorfman, Shames, & Kihlstrom, 1996). Finding a target digit in a matrix of characters likely involves just ordinary nonconscious processing. Sexing chickens at one day of age by touch seems anomalous, but then I know little about the biological characteristics of such creatures. The abilities of savants seem astonishing but, given the presence of systematic errors in many cases of calendrical savants, not extraordinary. There have, however, been some reports of extrasensory abilities among savants (Yewchuk, 1999). Can ordinary implicit cognition always account for intuition?

We saw in Part I, with the presentience studies and Nelson's global consciousness project, that there is some evidence for anomalous information transfer. Could the psyches of some savants be leaky so that they can draw on information from somewhere else? Could the chicken sexors be doing that? Was Grabhorn doing it? Was I? Am I? Could information from somewhere else be leaking in and out of our psyches all the time but getting lost in all the internally generated material? In other words, does implicit cognition have both ordinary and extraordinary components? Let us consider not just hunches, brief impressions, and tacit knowledge, but more extensive inner knowledge that apparently comes from somewhere else.

Channelling

Channelling refers to the ostensible transmission of information or energy through a person from unseen levels of reality to the everyday level. Channelling has also been called "mediumship" with the person who is doing the channelling known as either a "channeller" or a "medium" (Leskowitz, 2000, p. 164). At least some degree of dissociation is usually present with complete functional disconnection between a channeller's awareness and her behaviour in some instances. Channelled material can be brief or so extensive as to include lengthy expositions about the nature of reality. And, indeed, some channelled discourses have formed the backbone of various contemporary teachings as well as having become, in published form, best-selling books (cf. Klimo, 1987, Hastings, 1988; 1991; Brown, 1997; Decuypere, 1999; Leskowitz, 2000; Barušs, 1996; 2003).

What about the example of channelling, complete with equations, that I gave at the beginning of this chapter? John Barrow has noted that there has been a tendency on the part of some people to believe that the universe is a "vast cryptogram whose meaning could be unlocked by finding the right combination" and that he has received letters from the public claiming to have done just that through "strange jugglings of numbers" (Barrow, 1992, p. 6). On occasion, I, too, have had people present me with an ultimate equation or mathematical theory of everything that they have conjured up through years of some sort of deliberation.

Barrow's point, and mine, is that these imaginative mathematical productions are perhaps sincere but silly. And the equations for teleportation that a group of us channelled one evening have that same flavour. They appear to be unlocking secrets of the universe without actually making any sense. At least the metaphors for teleportation that we confabulated are more meaningful, even if only as science fiction. This is not to say that teleportation can or can not occur. We do not know yet the extent to which we will be able to eventually manipulate physical reality. But this channelled material, for all that it has a patina of legitimacy, appears to have little to do with reality.

In my experience, whatever the alleged source of channelled material, much of it is not particularly enlightened. In fairness, however, one needs to scrutinize it carefully before making judgments. And that applies to any mathematical content as well. After all, we do use mathematical expressions to convey our ideas about reality. We had an example of such productive use in Part I with Bell's Theorem and other examples will arise in Part III. There is nothing inherently wrong with doing so. Thus worthwhile channelled material could, in principle, include useful mathematical expressions. But the presence of mathematical expressions, irrespective of how complicated they may be, is, of course, no guarantee of truth.

In deriding channelled knowledge in general, however, I am not claiming that such material is always necessarily just the product of a channeller's own mind. Indeed, the danger, as I see it, is that there may exist entities whose presence we cannot apprehend through our physical senses but which nonetheless could intrude on us. Yes, nonconscious processing takes place, but at this point in time we cannot rule out the possibility that that processing might include elements of reality outside of our psyches. It is possible that the source of channelled material really is from somewhere else. And, if that is the case, then whoever or whatever is the source of that information may be less intelligent than we are. Furthermore, its motives may not be in our interests. And it may not be willing to disengage itself once it has latched on to us. Such potential characteristics of the source of channelling could pose a problem in cases where a person has relinquished volitional control over her body, feelings, and mind so that considerable psychological dissociation is present (Barušs, 1996; 2003).

In taking a critical stance toward channelling I am also not saying that it can never produce correct information. In a series of experiments at the University of Arizona, Gary Schwartz sought to determine the accuracy of knowledge provided by several mediums, among whom was John Edward, the host of a television show in which he ostensibly communicated with the dead (see Edward, 1998; 2001; 2003; for Edward's own account of his experiences). In the experiments, the mediums would produce information about a person, called a "sitter", apparently obtained from the deceased relatives and acquaintances of that person. In each case, the medium did not know the sitter, who sat behind the medium or was separated from the medium by a screen, and who either responded with yes and no answers to the medium's questions or remained silent. The information obtained by the mediums about the particulars of the deceased was compared with information provided by the sitters. Analyses of the data revealed that the mediums were providing statistically significantly more accurate information than would occur through guessing. For example, regarding information about the first sitter in the first experiment, the average accuracy for five mediums was 83% whereas the average accuracy for 68 undergraduate university students asked to guess the same information was 36%. The results could not be explained through fraud or error. In other words, whatever the source of the information and whatever the psychological mechanisms used by the mediums for obtaining it, correct information about the deceased was being provided by the mediums in this experiment (Schwartz, 2002, particularly p. 299).

Dynamics of Inner Knowledge

I have been giving examples of inner knowledge, such as intuitions and channelling, without specifying how such knowledge emerges in a person's awareness. I will try to address that now while giving additional examples of inner knowledge.

Experience, as it occurs for a person, consists of a sequence of thoughts, images, feelings, sensations, and such. In other words, our experience is a stream of consciousness, or simply "thinking", as we commonly refer to it, that goes on for us. By sampling people's experience at random times throughout the day, Eric Klinger and his colleagues have found a number of dimensions of thinking. In particular, thinking can be deliberate or spontaneous. On the one hand, thinking that is deliberate is thinking that is aimed at the attainment of specific goals and intentionally guarded against drifting from those goals through the wilful control of attention. On the other hand, spontaneous thinking consists of thoughts that occur without being volitionally directed. About one third of our thoughts are predominantly spontaneous (Klinger 1978; 1990; Klinger & Kroll-Mensing, 1995).

The occurrence of inner knowledge appears to be associated with spontaneous rather than deliberate thinking. Graham Wallas has given a famous characterization of problem solving in which illumination regarding the solution of a problem follows upon a period of incubation during which a person intentionally refrains from thinking about the problem to be solved (Wallas, 1926). Incubation itself appears to be a complex process whereby nonconscious mechanisms work toward a solution (Dorfman, Shames, & Kihlstrom, 1996) and does not appear to involve deliberate thinking about the problem to be solved. In a recent empirical study of intuitive experience, it was found that letting go, connecting with an object to be explored, and listening to the object, preceded an intuitive experience. Also in that study, it was found that intuitions could not be compelled to occur (Petitmengin-Peugeot, 1999). Thus inner knowledge appears to emerge into the experiential stream spontaneously.

An alternation of deliberate and spontaneous thinking can be used to try to evoke inner wisdom. This is often done through the use of guided imagery. For example, a person could deliberately imagine climbing a mountain until she reaches the peak. Once at the top, she could imagine a light in the distance that approaches her and resolves itself into the image of a wise person. In her imagination, she could engage the wise person in conversation (cf. Barušs, 2003; Ferrucci, 1982). It takes deliberate thinking to follow the directions of the exercise, but spontaneous thinking produces the image of the wise person and the wise person's answers to one's questions.

But, in the preceding example, do we need the picture of a wise being? For that matter, do we need the guided imagery? Indeed, we could just ask a question and then pay attention to the next image, thought, or feeling that arises in our minds. This appears to have been what Grabhorn did in the example given earlier, in which she wondered what to do about her low oil situation and had a feeling that she should take the next exit.

The elucidation of these dynamics raises an obvious question. When are spontaneous thoughts actually knowledge? After all, can the mind not apparently create a myriad of fictional variations on reality? And indeed, it seems to me that *the greatest problem associated with access to inner knowledge is the identification of that which is veridical or useful within the flux of the imagination*. It is the case, as revealed through empirical research, that daydreaming, consisting of spontaneous or fantastical thoughts, serves a number of necessary purposes within the cognitive economy of our minds, not the least of which is to remind us of our current concerns. So there are known benefits of spontaneous thinking (Klinger, 1990, whose definition of daydreaming I am using). But aside from clearly recognizable facts about our lives, how are we to know what is true? If we ask an imaginary wise being a question and butterflies take off, one after another, from his hand, then how is that an answer to our question?

What better way to answer that question itself than to use the very method being questioned. I did so during an eight day psychosynthesis workshop in which I participated. I had been asking questions and getting spontaneous answers. But I did not know if the answers were correct. As recorded in the journal entry for July 14, 1978, I asked another question: "How do I know the validity of the visions that are given to me as a result of my questions?" I got another image. A flower with a long stem and white head had grown up in a narrow flower bed outside the wall of a cathedral that I later identified as the church that I had attended as a teenager. There was a cubical frame around the flower, one side of which was formed by the wall of the cathedral. "The flower was myself and the cube was the cube of knowledge. The sides one after another were filling in with stained glass. But I knew that the top would not fill in because then the flower would be killed. The sides protect the flower from the side, but leave it open to above." I took this to mean that I could ascertain the validity of answers to questions pertaining to ordinary affairs but not those concerned with metaphysical matters. Indeed, that the flower stretched above the cube of knowledge suggested to me that there was inherent freedom in intuitive and mystical knowledge that could not be contained in "simple mental structures". I wrote in my diary that "I must reflect on the images and find what is meaningful in them regardless of their validity, and to choose to act on what is meaningful".

What I also came to realize was that an affirmative answer to the question about validity at that time in my life would have undermined my effort toward self-determination. After all, if I had thought that all questions had ready answers, then what need would I have had to try to understanding anything ever again? I could have become a "puppet of god" with the ironic consequence of curtailing my own process of self-transformation. Over the years, I have come to realize that the spontaneous productions of the mind can contain considerable wisdom, but I am no longer in danger of becoming inauthentic. I have made a point of strengthening my will so as to be able to act on the basis of my own understanding rather than being intimidated by transcendent imperatives.

Images that first arise spontaneously can be periodically revisited. I have done this several times over the years with the image of the flower. At one point it had become a young tree that had shattered the glass and destroyed part of the church wall. As it turned out, there was no church behind the wall. Years later, the tree had matured, there were some ruins still left of the church wall, but there was no glass to be found. The images coincided with my own sense that my will had become sufficiently strong so as to be able to incorporate inner knowledge without losing my integrity (Barušs, 1996).

I think that this example raises an important point, namely, that it may not be appropriate to have answers to metaphysical questions at a particular stage of self-transformation because the availability of such answers can undermine the process of self-transformation itself. At a later stage of development, more accurate metaknowledge about the validity of answers can be beneficial. This state of affairs requires patience on the part of anyone who feels uncomfortable about temporarily suspending judgments about the truth value of contentions about reality.

Varieties of Access to Inner Knowledge

Much of the time, spontaneous answers to questions do not appear to be particularly wise. One approach to this problem is to maintain that the person receiving the inner knowledge has contaminated the impressions with her ordinary thinking, hence rendering them useless. Let me situate this problem within one particular context, namely that of military remote viewing.

In 1972 Hal Puthoff, a physicist at what was then the Stanford Research Institute, was asked by representatives of the Central Intelligence Agency of the United States to conduct research into extrasensory perception (Puthoff, 1996). Because "extrasensory perception" had pejorative connotations within the scientific community, other expressions such as "remote perception" and "remote viewing" were introduced (Schnabel, 1997, p. 149). Puthoff ended up directing a program of research into remote viewing and related phenomena for 13 years with some striking results (Puthoff, 1996; see also Targ, 1996; May, 1996; Targ & Katra, 1998; Schnabel,1997; Graff, 2000; Mandelbaum, 2000; Morehouse, 1996.). Independent replication of research concerning remote perception was obtained at Princeton University as part of the Princeton Engineering Anomalies Research laboratory program (Nelson, Dunne, Dobyns, & Jahn, 1996; Jahn & Dunne, 1987).

My understanding is that the idea in remote viewing is to register the impressions about a target directly in the mind through nonsensory means without the interference of conscious thinking. Interpretive elaboration of a signal is called "analytic overlay" and analytic overlay is to be avoided in remote viewing lest it contaminate the quality of the information that is being received (cf. Brown, 1996). In other words, remote viewing is regarded as a process of perceiving a nonsensory signal with as little interference as possible from one's own understanding about the target being viewed.

But what if we were to proceed in the opposite manner? Rather than trying to set aside whatever we already think about a matter in question and allowing spontaneous thoughts to arise in our experiential stream, what if we were to engage in deliberate thinking by attending to that which we already know? That seems somewhat obvious and not at all intuitive, but that is actually the process of concentrative meditation. In concentrative meditation, a meditator chooses a seed, which can be an idea, an image, or some other mental construction, and directs attention to that seed for some period of time. In practice this usually means thinking thoughts that are semantically associated with a seed thought. One purpose of concentrative meditation is to persist in one's focus until previously unrealized aspects of the seed occur within one's experiential stream. More dramatic changes of consciousness could also occur, but we will leave discussion of those until Part III (Barušs, 1996; 2003).

There are two benefits to proceeding by cultivating deliberate rather than spontaneous thinking about a matter in question. The first is that the problem of interference by what one already knows is obviated. The second benefit is that there is full engagement of one's own volitional resources. What one already knows is a part of one's understanding, and any insights that occur during the process of concentrative meditation could also become integrated with one's understanding. In this way, instead of the functional dissociation required of remote viewing, a greater degree of integrity could be retained. I have not, however, seen any outcome studies comparing the safety and efficacy of proceeding by eliminating versus cultivating one's own understanding.

In concentrative meditation, insights can emerge through spontaneous thinking after a sustained period of deliberate thinking. But it seems to me, at least from my experience, that insights can also blend with intentional thoughts so that spontaneous thinking is superimposed on deliberate thinking. My experience sometimes has been that of having my knowledge supplemented. Rather than an alteration of thoughts that are volitionally directed with thoughts that are spontaneous, it seems to me that *understanding, as such, is a fundamental aspect of our experience into which contents can be introduced through different mechanisms*. These mechanisms could include controlled sequences of thoughts, the apparent results of skull-bound nonconscious computational processes, as well as intuitions emerging from somewhere else. This is not to say that the sources of specific information could not or should not be identified, but that they can serve the purpose of augmenting our understanding rather than functioning autonomously.

So we end up at the beginning in that we can simply seek to understand whatever it is that we want to know. However, our understanding could, and maybe always does, draw on extraordinary sources as part of implicit cognition. In effect, I think that the effort to isolate non-sensory perceptions can simply become a particular strategy in the service of understanding.

Interpretation of Symbols

I have already mentioned instances in which access to inner knowledge is mediated through figurative symbols. My image of the plant surrounded by glass is an example of that. The presence of such symbols is a serious roadblock to experimental investigation of inner knowledge given the multiple possible meanings that they can have. However, in spite of this difficulty, coming to grips with symbols is unavoidable in any serious discussion of access to inner knowledge.

Why the presence of allegories and metaphors rather than direct depictions of the substance of inner knowledge? Is it because the truths that they express cannot be entirely conveyed in literal terms (Jung, 1921/1971)? Or do they protect inner knowledge from those who are unprepared to use it responsibly? Even some of the military remote viewers apparently found that anomalous input was frequently symbolic (Mandelbaum, 2000). I do not know why inner knowledge is often portrayed figuratively rather than literally, particularly in cases where it would appear that literal depictions would be more direct. But such symbolization appears to be a natural capability of our minds, as we can see, for example, from the nature of hypnagogic imagery.

Hypnagogic imagery is imagery that occurs as we fall asleep. It can take place in any of the sense modalities, although visual images predominate. What is interesting is that hypnagogic images can be autosymbolic in that they can be figurative depictions of immediately preceding thoughts about one's mental state or one's bodily sensations (Mavromatis, 1987). In other words, thoughts about one's thinking and about one's somatic feelings are automatically translated into images that occur within one's experiential stream.

Let me give some examples of the autosymbolic nature of hypnagogic imagery. At one point as I was falling asleep I was thinking about how to get past the reflective aspects of the mind to a deeper reality underneath, when I spontaneously had the visual image of a mirrored wall and a door opening beyond the mirrors. In this case the visual image of a door among the mirrors was a depiction of my musings about going past the ordinary aspects of the mind. Another example is one from a moment of reverie during which I may or may not have been falling asleep. I was obsessively worrying about something when I had the clear image of a fly buzzing around a room. I realized that my mind was attached to the fly and that I could shift my attention away from the fly to the room as a whole. With the second example I am already suggesting that symbolic imagery need not occur just during the hypnagogic state and that it may be richer in content than the simple reflection of preceding thoughts and sensations. In the second example, there is an implicit solution to my worrying, namely, to refocus my attention. But that just brings us back to the notion that there can be figurative expressions of inner knowledge.

The interpretation of symbols depends upon the ability to match the pattern of characteristics of the symbolic images with events from one's life. There can be both culturally common interpretations for images as well as interpretations that are specific to an individual based on her history and understanding of reality. Variations on the same symbols with related interpretations can occur for an individual over time. Although there could be primary interpretations for particular symbols, other interpretations could also fit. It is as though a language of communication develops over time with nonconscious parts of oneself (Barušs, 2003). Some examples of interpretations of symbols have already been given, and others will follow later.

Alterations of Consciousness

We have been discussing the dynamics of inner knowledge primarily in the context of the ordinary waking state of consciousness, although concentrative meditation and hypnagogic imagery both involve alterations of consciousness. In concentrative meditation, the meditator becomes absorbed in the subject matter of meditation so as to be oblivious to distractions in the environment around her and hypnagogic imagery occurs as one falls into the altered state of sleep. The question is whether access to inner knowledge can be improved when consciousness is altered. The answer is that I think so, based on my own experience and the proclivity for anomalous phenomena to occur during alterations of consciousness (Krippner & George, 1986; Barušs, 2003). Let us more carefully examine inner knowledge in the context of alterations of consciousness.

Over the years I have often heard someone claim that she is putting herself into a "trance" in order to access inner knowledge. I have sometimes used that expression myself. It sounds as though something amazing were happening, but what exactly is taking place? Well, there is no single identifiable state of trance, and, indeed, the word "trance" has had various meanings. In general, the term refers to alterations of consciousness characterized by the apparent presence of awareness accompanied by involuntary behaviour and diminished responsiveness to environmental events (cf. Pekala & Kumar, 2000). In other words, a person in trance is someone who is absorbed in an apparently mindless activity to the point of being resistant to interruption by distractions. In effect, what I think a person usually does who claims to be going into a trance is to redirect attention away from sensory stimuli toward internal events and possibly shift toward more spontaneous thinking. This could more simply be referred to as daydreaming (Barušs, 2003; cf. Klinger, 1990). But there could also be more dramatic processes involved in trance. In particular, a person could experience some degree of dissociation.

Dissociation, as we have been using that word, refers to functional disconnections of the psyche (cf. MacMartin & Yarmey, 1999; Cann & Harris, 2003). In that sense, we experience dissociation all the time. For example, we might engage in a thoughtful conversation while automatically going about the business of driving a car. That is clearly an innocuous disconnection. However, there are degrees of dissociation. At the most extreme end, a person can experience the symptoms of dissociative identity disorder, whereby, as we saw from Part I, several distinct personalities appear to function independently through a person's body. Channelling also involves dissociation insofar as the process of channelling is disconnected from a person's own volition. Hence, trance, in the context of inner knowledge, can entail a considerable degree of dissociation.

The reader may be wondering whether channelling is just dissociative identity disorder. The answer is no. There are considerable differences between channelling and dissociative identity disorder. For example, finding objects to be missing, handwriting changes, amnesia for childhood events, flashbacks, auditory hallucinations, and referring to oneself in the plural are all symptomatic of dissociative identity disorder but not channelling (Hughes, 1992). It is possible, however, that channelling occurs during dissociative identity disorder, given that some people with dissociative identity disorder have personalities that have been identified as demons or dead relatives, as noted already in Part I. Apparently greater than normal transliminality in dissociative identity disorder could result in greater ingress of anomalous material, although such material may or may not constitute inner knowledge.

Trance is also the term used to designate someone who has been hypnotized. Hypnotic behaviour is characterized by responsiveness to suggestions, which can include, for example, making or inhibiting suggested motor movements, decreased feelings of pain, increased forgetfulness, and changes of identity. Can hypnotic trance facilitate access to inner knowledge? There has been considerable confusion about the nature of hypnosis in the academic literature with strong polarization of the research community. However, some points of clarification can be made.

The first point that can be made about hypnosis is that a hypnotic induction, used to bring about a hypnotic state, is just a particular guided imagery exercise labelled as hypnosis. Second, hypnosis really is a substantially modified state of consciousness for those who are particularly susceptible to it. Third, there appear to be three separate psychological factors that can contribute to a hypnotic trance. The first factor is simply a positive attitude toward the experience of hypnosis in which the hypnotic subject strives to comply with a hypnotist's suggestions using her ordinary cognitive abilities. The second factor is fantasy proneness. There are some people whose imaginations are so vivid that their fantasy life appears to them to be as real as the ordinary sensory world. The third factor is dissociation, which can lead to phenomena that are attributed to hypnosis (Barušs, 2003; for the tripartite theory of hypnosis, see Barber, 1999).

The most obvious application of hypnosis as a means of access to inner knowledge is through its use in regression. The idea is that a state of trance could facilitate a person's ability to recall previous autobiographical events. For example, a person could be hypnotized and led to remember what happened to her during childhood, during childbirth, or during a past life. Well, such accounts can certainly be collected. The question is whether they represent actual events or false memories. There has been considerable controversy over the actuality of memories obtained through regression even without considering cases of remembered childbirth or past lives. In general, experimental studies have shown that hypnotically facilitated memory recall can produce more memories, but that much of these are not actual memories, but confabulations. The problem is that hypnosis increases the confidence that these really are memories (cf. Schacter, 1995; Frankel & Covino, 1997). The exception may be for the recall of traumatic memories, which may be encoded differently, so that the central features of such memories are retained intact but possibly disconnected from ordinary awareness (cf. van der Kolk, 1996; McFarlane & van der Kolk, 1996; Krystal, Bennett, Bremner, Southwick, & Charney, 1996; Brown, Scheflin, & Hammond, 1998; Nadel & Jacobs, 1998).

Despite the problems, there have been claims of knowledge acquisition through hypnotic regression. In a study of ten mother-child pairs in which the mothers and children were hypnotized individually, it was found that the children could recall events that had occurred during their birth. There were, however, methodological problems with the study (Chamberlain, 1986). Regression to past lives is even more controversial, although there is some evidence from children's remembered past lives for the possible occurrence of actual past lives (Mills & Lynn, 2000; Stevenson, 1997a; 1997b.). There have been claims, not only that past lives exist, but that knowledge of past-life events can help a person deal with physical or psychological health problems in a current lifetime. For instance, becoming aware of having fallen and dying of a head injury in a previous lifetime could lead to remission of recurrent headaches in this lifetime (Freedman, 2002). Although some clinicians feel, on the basis of their clinical experience, that past-life regression has been therapeutically effective, there has been almost no systematic research to investigate that claim. In a rare experimental study, one of my students, Kellye Woods, using a guided imagery technique, regressed 24 healthy student participants in the psychology laboratory at King's University College in order to find the possible psychological benefits of past-life regression. She found no substantial effects on psychological well-being of the presence of past-life imagery (Woods & Barušs, 2004).

Dreaming

Dreaming is an altered state of consciousness that is, yet again, controversial. There are those who have maintained that dreams do not mean anything. Or at least nothing of any substantial significance (e.g., Hobson & McCarley, 1977; Hobson, 1988; 1990; Hobson, Pace-Schott, & Stickgold, 2000). For those who do think that dreams are meaningful, there is a variety of ways in which they are thought to be meaningful and numerous schemes for interpreting them (Barušs, 2003). Some of my dreams are certainly meaningful to me and have provided me with many useful insights.

The realization that actual knowledge could sometimes be present in my dreams did not occur overnight, but over the course of several decades. This is consistent with the contention of the British mystic, Douglas Baker, who has said that he had to pay attention to his dreams for 25 years before he could synthesize an effective set of symbols for communicating with a wiser part of himself (Baker, 1977). I do not think that it needs to take that long for a person to be able to learn from her dreams, but such time-lines do attest to the need for patience.

I think that some of the confusion concerning dreams has been the result of thinking that dreams are fixed programs that run in our minds, insensitive to how we approach them. I remember being struck upon reading Frances Vaughan's statement that "it is common knowledge that people in Jungian analysis have Jungian dreams, while those in Freudian analysis have Freudian dreams" (Vaughan, 1979, p. 104). Of course Freudian psychotherapists would interpret the same dream differently than Jungian psychotherapists, but to actually dream different symbols in dreams seemed unreasonable to me. And yet, if anything, Vaughan had understated the case. Our minds and, in particular, our dreams, are more malleable than we think. Let me give several examples of the degree to which dreams can respond to our intentions.

I have been teaching a popular course in altered states of consciousness to students from across the university. Among the topics covered in class has been the induction of lucid dreams. Lucid dreams, dreams in which a person knows that she is dreaming, turn out to have numerous benefits for a person, such as providing her with an opportunity to practice waking-life behaviours. There are various techniques that can be used for increasing the likelihood of lucidity. In particular, I have discussed a method that was developed by Stephen LaBerge. As a person falls asleep, she alternates the following two activities: she deliberately intends to realize when she has a dream that she is dreaming and she imagines herself becoming lucid in a previous dream that she has had. I have also mentioned the notion of analysing, upon awakening, one's dreams for unrealistic events. Events that are physically unlikely can serve as clues to a person that she is dreaming (LaBerge &

Rheingold, 1990). For those with difficulty remembering dreams, I have pointed out that, immediately upon awakening, one could try to trace one's thoughts back to the moment just before awakening (cf. Wren-Lewis, 1988.).

One day a student showed up in my office who was visibly concerned about something that had happened to him. He had not been a prolific dreamer, nor had his dreams been particularly interesting. However, as a consequence of my lecture, he had decided to try to induce lucid dreaming in himself. He had developed his own "mantra" consisting of the expression "I will dream real, I will become lucid, I will remember everything". He repeated this expression after going to bed until he fell asleep while at the same time imagining becoming lucid in the most recent dream that he had had, which had been some time considerably in the past. In the morning, immediately upon awakening, he would try to trace back his thoughts to the last dream that he had had. He did this for about two months. Obviously the "mantra" was not what LaBerge had suggested, but could be considered a somewhat creative variation on what he had proposed.

For the first three weeks, nothing happened for the student. But then, he said, he started waking up in the night right after having had a dream. He would be wide awake, recalling a "vivid, often colourful narrative that was guided by an interesting plot". Almost immediately afterward, he would fall back asleep. This started to occur two or three times a night. But it did not stop there. After about one and a half months, he said that he no longer needed to trace back his thoughts to the point of awakening, but would find himself in strange situations in the morning acting out in real life the end of his dream as though he were still dreaming. When this started to happen consistently, he decided that he should stop trying to induce lucid dreams. However, these waking episodes continued for another two weeks. For example, at the end of one dream, the dreamer was in a grocery store packing loaves of bread into grocery bags. The student said that that day, in real life, when he got to the library, he reached into his school bag for a book and pulled out his clock radio. He realized that he had woken up that morning and put the clock radio into his school bag thinking that he was still packing groceries.

I find this account interesting, in that the pre-sleep suggestion to "dream real" appears to have been interpreted literally by the dreamer's psyche. That is to say, the dream became physical reality. Ironically, the originally intended behaviour, the induction of lucidity, never did occur. Rather, somewhat the opposite happened, in that the dream events intruded into ordinary reality for some period of time after awakening.

There is another way in which dreams can be responsive. The events in dreams may occur in such a way as to try to direct the dreamer's attention toward something. It is as though there were an intelligence pointing toward particular dream images. The first time I explicitly suspected that that could occur in dreams was during a study of dream telepathy conducted by one of my students.

Two of the students in one of my classes said that they regularly found themselves in each others dreams at night. In other words, they would ostensibly dream the same dream, each from her own point of view. There has been almost no research concerning such dream sharing, probably because the possibility of sharing dreams is considered impossible from a conventional point of view. I suggested to one of them, Gillian Johnson, that she carry out an experiment in dream telepathy, whereby she would visualize images before falling asleep and then see if they would show up in the dreams of the other student. A third student, who did not believe that dream telepathy was possible, agreed to participate as a control, in that we would see if any of the dream images showed up in his dreams.

We set up the experiment as follows. I gave Johnson seven sealed envelopes, each of which contained a slip of paper with five neutral items or events written on it that were to serve as the targets. Each night of the experiment, at her own house, she would open one of the envelopes and make up a story involving the five targets that would also include the presence of negative emotions, since previous research had shown that negative emotions facilitated dream telepathy. She would also draw a picture illustrating the story. All that the two students who functioned as the receivers had to do was to record on audiocassette recorders any dreams that they had had during the night after sleeping in their own homes. The participants were not to communicate with each other about their dreams until the data had been gathered.

Now, because we had not developed a numerical method of scoring we could not assign probability levels to the results and because we had not implemented safeguards against deception the outcome of this study would not be accepted by the scientific community. Thus, the correspondence that we did find between targets and dream images is speculative only, but what is noteworthy is the directedness of the recipients' dream events toward the targets. For example, one set of targets consisted of a broken watch, the former Soviet leader Mikhail Gorbachev, a divorce, a hockey game, and shopping at the Bay— "the Bay" being a department store chain in Canada. The story created by Johnson consisted of coming home from shopping at the Bay after buying a book about Gorbachev and finding her house to be dark. She tries to turn on the lights but nothing happens. She goes into the kitchen where she is accosted by a man who threatens her if she tries to divorce him. They struggle. She picks up a hockey stick and beats him over the head and shoulders with it. He throws her against the patio door, shattering the glass, breaking her watch and cutting her right hand in the process. The drawing accompanying the story included a portrait of Gorbachev complete with the large dark blemish on his head. This was the story that Johnson made up before falling asleep.

One of the participants in the study dreamt that she was attacked by wolves that had crashed through a window into a building. The wolves chewed her hand (later clarified as being the right hand) to the point where it was bleeding. A wolf lunged for the dreamer's face. The participant later added that she had smashed a wolf over the head with a "metal stick that police use". At one point in the dream she had found it necessary to turn off the lights (Johnson, 1990).

Johnson and I were struck by the apparent agency of the wolves in the dream. The wolves at times appeared to be acting like messengers trying to draw the dreamer's attention to the targets directly or through evoking particular dream actions. This is suggested by the wolves chewing the hand until it bled. To understand how odd the correspondence is between the story created by Johnson and the actions of the wolves in the participant's dream, the reader can try to remember the last time that she had a dream in which her hand was bleeding.

OK. I may be extrapolating a bit too far from the data. This example is not evidential but, as I said, it did lead me to suspect that dreams could be structured in such a way as to draw attention to particular dream contents. I will give a better example in a moment, but the potential responsiveness of dreams to our intentions does lead to an obvious question concerning inner knowledge. If inner knowledge were to be available in dreams, could we receive specific information on request? In other words, can we apply the alteration of deliberate and spontaneous thinking to dreams, where the deliberate thinking consists of our query and the spontaneous thinking corresponds to a dream? This is the subject matter of dream incubation, about which there has been a little bit of research.

Dream Incubation

A prototypical experiment in dream incubation is to give someone a brain teaser and see if she can come up with the answer in a dream. Morton Schatzman gave the following brain teaser to a fellow physician: What are two words in the English language that begin and end with "he"? The physician subsequently had a dream in which the dream character falls over from chest pain and is rushed to a hospital with people laughing "hee, hee, hee". The dream character is asked by a doctor what his condition is called, in fact, what it could be called in plain language, until the dream character answers that it could be called anything, even heartache. But there is still some pain. The dream character is to go to a word specialist. Schatzman appears in the dream and tells the dreamer that he had told him that there were two things wrong with him, and that he must learn to juggle words and pains. The dream character says that riddles give him headaches. And with that, the pain is completely gone. The words are, of course, "heartache" and "headache". More clearly than was the case in the dream with the wolves, there appears to be an agency operating in the dream directing the dreamer's attention to the correct solutions of the brain teaser (Schatzman, 1983a).

Dream incubation can range from simply asking a question at some point with the intention of getting an answer in a subsequent dream to formal dream incubation rituals. Henry Reed developed a dream incubation procedure based on the rationale that a desirable psychological condition could arise from alignment with the symbolism of a ritual. In other words, the idea was that meaningful enactment of a particular pattern of behaviour could induce desired psychological states. As part of a ritual that he developed, based on historical precedents of dream incubation, Reed asked participants to choose and draw symbols of a "sacred place" and a "revered benefactor". A sacred place was to be a place that evoked feelings of reverence and nurturance for a participant, while a revered benefactor was an esteemed and inspiring person, either actual or fantasized. Prior to going to bed in a tent that was used as a sleep sanctuary, Reed engaged the participant so as to evoke expressions of psychological material relevant to the problem for which a dream solution was being sought and had the participant play the roles of the revered benefactor and the sacred place. Finally, Reed guided the participant through a "presleep reverie" that included the symbols of the sacred place and revered benefactor. Its purpose was to help the participant to trust the automatic processing that occurs in dreams as well as to discharge more superficial material so as to free the sleep activity for engagement with deeper psychological strata. In the morning, Reed would return to the tent and ask the participant to recount the dream and its possible meanings, and have the participant produce a written account of her entire experience with the dream incubation event (Reed, 1976; quotations from p. 60).

Reed found that participants often had meaningful dreams that addressed their concerns. In some cases, a dream's value lay not in the interpretation of its symbols but directly in the experiences with which it provided the participant. For example, one man, who felt inhibited in the work that he did as a result of self-criticism, had a cathartic dream about interactions with significant people from his past which resulted in his awakening with a renewed capacity for his work. Sometimes participants had visions during the night that seemed to them to be qualitatively different from dreams. For example, one woman said that she had apparently awoken during the night to find that a strong wind had blown away the tent. She had ended up interacting with an old woman and reading a tablet that had purportedly contained information about her "past and future lives". At the abrupt conclusion of the vision she had found herself back in the tent "as if she had awakened from a dream" (Reed, 1976; quotations from p. 66). Reed's induction is a particularly comprehensive one that can apparently result in some dramatic experiences for the dreamers.

Less onerous procedures for the incubation of dreams could include the following ingredients. The dreamer could make a written statement of the problem in her journal and place it beside her bed along with a writing implement. Objects that are relevant to the problem could be arranged near the sleeper. After going to bed, the dreamer could review the problem, visualize having dreamt a solution, and suggest to herself that she will dream about the problem as she falls asleep. Upon awakening, she can lie in bed recalling any dreams that she had before writing them down (cf. Barrett, 2001).

Deirdre Barrett asked 76 college students in her class to incubate a dream within one week in order to solve a personal problem that each of them had. They were to use a dream incubation technique that they had learned in class (that of William Dement, 1972/1978). As judged subsequently by the objective evaluations of two raters, 67% of the students had dreams that were on topic and 33% had dreams that presented solutions to their problems. For example, one student could not decide which graduate program to enter. She dreamt of flying over a map of the United States when the plane experienced engine trouble and needed to land. The dreamer asked about landing in her home state and was told that it was dangerous to do so. The safe place, pointed out by a light on the map, lay further west. Upon awakening, the student realized that it was more important for her to get away from home than concern herself with the particular program she entered. This was a dream that addressed the stated problem and presented a solution as judged by both the participant and the two raters (Barrett, 1993).

I thought I would experiment with a couple of the brain teasers provided by Schatzman. The first of those was the following: *"The numbers 8, 5, 4, 9, 1, 7, 6, 3, 2 form a sequence. How are these numbers ordered?"* And the second was: *"What is the missing letter? H, Z, X, O, I, S"* (Schatzman, 1983b, p. 417). I tried for a while to figure these out and could not, so I lay down to take a nap, with the intention of solving the first of the two.

I held the intention to solve the first of the two brain teasers in my mind as I fell asleep. Upon awakening, the only incident from my dreams that I could clearly remember was a feeling of becoming completely irrational while giving a talk in public. In the dream, I decided to keep going with the talk despite the loss of coherence. In my waking state, the associations that immediately came to mind were those of light-headedness and a sense of elevation. It seemed to me that I had had a bird's eye view of myself from outside my body. Maybe the dream pertained to the second of the brain teasers, I thought, in that the light-headedness and elevation could have represented the effects of helium, a word that starts with the letter "H". Perhaps these were the inert gases. The letter "X" could stand for "xenon." But the rest of the letters did not fit. I went back to the first brain teaser and, using the idea of elevation, arranged the numbers in the form of a descending staircase. And then, as I thought about the problem, the answer became clear. In retrospect, given that my dream occurred during a short nap, it was likely an autosymbolic hypnagogic image representing the process of relinquishing rationality in order to try to draw on implicit cognitive resources, rather than a direct clue to the solution of the brain teaser.

That night as I was falling asleep I clearly intended to solve the second of the two brain teasers. During the night I dreamt over and over again that I knew the answer and that the answer was obvious. But I remember being frustrated at one point during my dreams thinking that it was no use realizing that the answer is obvious without actually knowing the answer. When I woke up, the only other thing I could remember from my dreams was seeing the sign of a grocery store in Toronto that said "Dominion." There really was such a grocery store chain in Canada. But what was interesting about the image were the characteristics of the sign. First, the letters were three dimensional and sat on a low stone structure in front of the store. Second, the first letter of the word was much larger and more elaborate than the rest of the letters. Third, upon reflection, the first letter was actually "G" and not "D" even though the word was "Dominion". This last effect is analogous to the superposition of states in quantum mechanics, in that the first letter of the word has two values rather than just one. I felt that the biggest clue from the dream was that the answer had to be obvious, so I looked for something obvious. Second, I thought that the key must have something to do with the letters themselves, with changing them somehow. I realized, after all, that there are few words that begin with "X", if the letters were to be the initial letters of words. As I thought about these constraints, I manipulated the letters in my mind and realized what the solution was. Even though my dreams were more helpful for the second brain teaser than the first, ultimately it was through the process of thinking logically and creatively that I was able to solve them. However, in doing so, *I incorporated the spontaneous productions of my mind, both from my dreams as well as my waking state, within the process of my own understanding.* I have left out the actual answers to the brain teasers in case readers would like to try them for themselves.

Well, if dreams can at least sometimes respond to queries, are there limits to the types of questions that could be posed? Can dreams inform us about our health, for example? And indeed, in Barrett's study, two of the students posed medical problems. In both of those cases the students had dreams that addressed and presented solutions as judged by objective ratings of the dreams. However, it was not made clear by Barrett whether the proposed solutions actually resolved the medical problems (Barrett, 1993).

Dreams About One's Health

Health problems and their apparent solutions, whether incubated or not, have been known to present themselves in dreams. In one study, dreams were obtained from patients who were to undergo cardiac catheterization in order to look for heart disease. It was found that the greater the number of references in their dreams to death and separation, for men and women respectively, "the more severe the heart disease" (Smith, 1990, p. 227). In another case, Wanda Burch used her dreams to negotiate her way through breast cancer. Among her dreams were recurrent dreams of a dance hall of the dead in which she danced toward a wooden door that represented her death, dreams of her father shouting at her that she had malignant breast cancer in spite of her doctors' assurances to the contrary, dream images of the cancer that turned out to be correct visual depictions of it, and a dream in which she killed a large bat and poisoned a whole host of small ones representing her chemotherapy (Burch, 2003). Let me give an example of a health-related dream from my experience.

I was living for a while in a small town in Eastern Canada. As recorded in my journal on January 16, 1994, toward the end of a dream before awakening, I was going down a hill back to civilization when I encountered large bear footprints. I started running down the hill. I came across some scaffolding where a forest ranger lived. I called him. He made me caress the teeth on the left-hand side of a bear's mouth with my right hand. I was still afraid, but he said that he would come with me to the town. I woke up from the dream.

On January 17, 1994, I remarked in my journal that I had a painful lump in the left side of my mouth. It was on January 18, 1994, that I associated the dream imagery of January 16 with the lump in my mouth and the ranger in the dream with Dr. Dave, a physician who raised goats on a farm in the highlands. I noted that I was still afraid of the bear even though, in the dream, the bear had been benign. If I were to take the dream imagery as referring to my current physical concern, then the dream had seemed to reassure me that there was nothing to worry about.

On January 19, I went to see Dr. Dave. Rather than calming my fears, he sent me immediately to a dentist who had the capability of taking a proper x-ray. The dentist raised my alarm even further. Upon taking the x-ray she said that the lump had nothing to do with any of my teeth and sent me to a specialist in a slightly larger town who had the capability of performing a biopsy. I was nervous. This was not looking good. Although they were not using the c-word, the medics clearly suspected cancer. However, when I saw the specialist the following day, he said that the lump was the result of irritation caused by breaking a small bone in my skull and was nothing to be concerned about.[11]

Bears symbolically represent fears for me. At first in my dream there were bear footprints, then a ranger had me actually caress a bear's teeth. But despite the close encounter, the bear is benign. There is fear, but no actual danger. In real life, I was afraid that the lump could be cancer. My initial fear was represented by the bear's footprints. The forest ranger I called upon in the dream was Dr. Dave. In the dream the ranger had me caress the bear's teeth, the site of my actual physical concern, thereby escalating my fear. In real life, my physician sent me for further medical examinations causing me to become more afraid. But in the dream, the bear was harmless; in real life, the lump was just a benign irritation. Thus it turned out in this case that there had been a correspondence between the dream events and the events in my life.

The Vagaries of Inner Knowledge

Much of the time I do not know why specific dream contents appear in my dreams, but I do not find them to be meaningful. I can recognize events in distorted form from the previous day, or images that are expressions of my wishes and fears, or narratives that depict what I think is happening. But there are dream events that arise out of the jumble of mental furniture that seem to belong to a more rarefied cognitive domain. Sometimes a dream strikes me as being such a dream and then I pay particular attention to it.

Just as the savants do not know how they know, so neither do I know exactly what it is about some dreams that indicates that they may contain real wisdom. I think that writing out dreams, identifying what they could mean, and then making predictions based on those meanings, can be used in order to find which dream images could correspond with which life events. I do not know that there is a short cut to such experimental work given that dream symbols are bound to be at least partially unique for each individual. But in that way, the identification of meaningful dreams could be possible.

Nonetheless, in reflecting on the matter, I have come up with a list of some of the characteristics of dreams that are meaningful for me. First, such dreams usually occur toward the morning, and I often wake up directly from them. Second, they are fairly short and to the point, rather than long with a meandering narrative structure. Related to this, there is a clarity or decisiveness about the dream events. It is as though they proceed without hesitation. Third, meaningful dreams are usually figurative rather than literal, but the symbolism is fairly transparent. That is to say, the dream images are familiar symbols that have had particular meaningful associations for me in the past.

There are also aspects of dreams that cannot be used to rule out their meaningfulness. One of those is the presence of the previous day's events in a dream. Dream research has shown that a previous day's events or environmental stimuli present during sleep are often incorporated in dreams. However, these events are not necessarily just replayed in dreams, but are often transformed in complex ways (De Koninck, 2000; Strauch & Meier, 1992/1996). One way to think of the previous day's events is to think of them as the building materials of the mind that are available for use by the dream architect to create a message. The point is that there can be instructive rearrangements of these materials.

Dreams that are meaningful can be brutally honest by confronting us with who we really are or drawing attention to events that we would prefer not to encounter (Stevens, 1995; Ullman, 1999). As Montague Ullman has said, there is an "ethical aperture" that operates at the core of our psyches becoming more apparent when we are asleep, just as a camera aperture opens wider in the dark. What is revealed through opening the ethical aperture is a better picture of who we are and the circumstances surrounding our lives so that we are in a better position to make moral judgments. However, according to Ullman, dreams can show us our moral choices but cannot enforce them (Ullman, 1999; quotation from p. 101).

In my experience, in straightforward situations, such as which publisher might be willing to publish one of my books, dreams can give unambiguous pictures. However, in more complex situations, such as those involving human relationships, the pictures can become a mosaic of different facets of the situation, some of which are desirable and others of which are not. No preferred course of action may be evident. Rather, I find that I have to make judgments based on my understanding—albeit understanding that has been augmented by knowledge that I would not have had were it not for the broadened perspective of my dreams.

In providing a broadened perspective, the dream architect does not appear to care whether or not we are supposed to have access through any conventional means to the information that is revealed in dreams. Such information can be displaced in space and time. That this is possible is suggested by the results of the remote perception and presentience experiments mentioned earlier in this book. But there has also been some direct evidence for the existence of meaningful coincidences between the contents of dreams and future events, as seen from the following precognitive dreaming experiment.

Malcolm Bessent, who was known for having had anomalous experiences, participated in a study in which the contents of his dreams the night before viewing an audio-visual slide show were compared to the contents of his dreams on the night after the show with regard to the degree to which the dreams matched the show's contents. This protocol was repeated for a total of 8 different showings. It was found that on 7 occasions, the preceding nights' dreams better matched the shows' contents than the following nights' dreams. For example, on one prior night, Bessent dreamt about the colour blue, water, and birds, and explicitly said that he thought that the next slide show would be about birds. The slide show actually was about birds. The subsequent night, Bessent dreamt about a secretary having dinner in a restaurant. The dream of the night before the slide show clearly was a better match to the show's contents than the dream on the night afterwards (Ullman & Krippner, 1973).

There is something else to which I have alluded several times in this book that is important to address explicitly, namely, the importance of authenticity. As a person becomes aware of the possibility of access to inner knowledge, there can be a temptation to regard inner productions of the mind as authoritarian edicts that need to be obeyed. This could be particularly the case if hunches are conceptualized as promptings from a higher self. After all, only a fool would defy the guidance provided by her own soul. But, as I have tried to illustrate, hunches are not infallible. Far from it. Much of the time they turn out to be fictional. And, ironically, one ends up running fool's errands.

Suppose, however, that one knows for certain that one's inner knowledge is infallible. Then, as I have already argued from my own case, there can be an even greater danger, in that one can be tempted to become a puppet of god. In such a scenario, a person relinquishes control of her own life in favour of guidance. Healthy psychological development ceases. In order for constructive self-transformation to take place, I believe that a person needs to engage all of her resources, including her capacity for rational thinking, creativity, and self-determination.

Inauthenticity is characterized by compliance with the expectations of others; in the case of inner knowledge, internal others. *Authenticity is the effort to act on the basis of one's own understanding.* Hence my continual emphasis on the importance of understanding as a fundamental feature of the psyche. And my emphasis on the exercise of the will (Barušs, 1996).

From the discussion in this section, inner knowledge sometimes functions precisely by releasing one's own interpretation of reality. This was represented by the irrationality I experienced in the dream whose intention was to solve a brain teaser. But any insights acquired through dissociated mechanisms can be brought back into the context of one's own understanding and self-determination. Opening up is important, but needs to occur, I think, from a position of psychological strength.

Access to Inner Knowledge

We started out by noting that materialism is dead and then considering the importance of self-transformation as a means of seeking answers to existential questions. Rather than abandoning science we reoriented ourselves to its essence, which consists of the synthetic development of understanding. And, indeed, we surmised that authentic science is consistent with the quest religious orientation and, hence, can be a form of spiritual aspiration.

In this part of the book we looked at an example of channelling and then some examples of hunches. Are the hunches just coincidences or is there something extraordinary about them? Or are they extraordinary coincidences? We considered chicken sexors and calendrical savants; messages from the dead; intrusions of dreaming into waking and waking into dreaming. There is an alteration of deliberate and spontaneous thinking during the ordinary waking state that can be carried over into alterations of consciousness such as hypnosis, meditation, and dreaming. In particular we can try to incubate dreams in order to answer questions. And we considered a number of examples of dreams in which knowledge was manifested, in some cases more evidently than in others.

Conventional explanations for correct inner knowledge are given in terms of implicit cognition, the operation of automatic psychological processes of which we are not aware. But what happens during implicit cognition? Is implicit cognition always just physical computation? We have seen some evidence to suggest that our understanding could be influenced by extrasensory and superconscious sources. Yes, channelling can yield silly teleportation equations, but it can also result in correct information about those who are deceased. Yes, our imagination will confabulate wonderful or frightening fantasies in answers to our questions but, at times, will also provide deeply significant insights into the nature of reality. Yes, dreams can be a jumble of imaginal and narrative snippets but they can also reveal something meaningful about the dreamer's life or about objective physical events. *Implicit cognition need not always involve just mundane computational processes but can apparently draw on extraordinary material for its noetic content.* And thus our understanding is sometimes augmented by material that is not just available through ordinary means.

Thus far we have considered inner knowledge that percolates into our awareness, as it were, there to be integrated with the totality of our understanding of reality. But there have also been reports of more dramatic forms of inner knowing, namely, reports of transcendent states of consciousness characterized by enlightenment and the resolution of existential problems. In the next part, let us turn to seeking transcendence by using science as a spiritual practice.

Part III

Seeking Transcendence

John Wren-Lewis, a graduate in applied mathematics, was 60 years old in 1983 when he was travelling with his wife Ann Faraday in Thailand.[12] They were getting on a bus when they were offered some toffees. Wren-Lewis ate the one given to him while Faraday showed more sense and did not eat hers. It turned out that the candy had been laced with drugs, probably morphine—so much so, in fact, that Wren-Lewis went into a coma.

Several hours after regaining consciousness, Wren-Lewis began to wonder why it was that the hospital room in which he found himself seemed to be so beautiful. He had been told that he had been poisoned but assumed that the drug effects would have worn off by then. Perhaps, he thought, he had had a near-death experience, and took himself back in his mind to the moment just before he had awakened. What he found was a "dazzling darkness" underlying his ordinary consciousness that was so palpable that it seemed as though the back of his head were exposed to the infinite reaches of space. Everything within his experience appeared to be sharp but distant as it would have appeared upon looking through the wrong end of a telescope. He felt that "some kind of brain-cataract [had been] removed, making unobscured perception possible for the first time" (Wren-Lewis, 1994, p. 109). His sense was that he *was* each of the things that he perceived at the same time that he was the living darkness out of which, moment-by-moment, all things were being brought into manifestation. Affectively there was a deep sense of satisfaction, peace, love, bliss, and joy (Wren-Lewis, 1988; 1991; 1994).

In the days that followed, Wren-Lewis found that his transcendent state of consciousness would lapse from time to time. However, it would return as soon as he identified that it was missing. He came to regard his endarkened condition, not as an alteration of consciousness, but rather as his baseline state of being. His endarkenment became the natural grounded state in which everything was truly the way it is, whereas his previous ordinary state of consciousness came to be regarded as the "real alteration" in that it had been a "clouded condition" in which he had had the impression that he was a "separate individual entity over against everything else" (Wren-Lewis, 1991, p. 6).

Wren-Lewis has advanced the hypothesis that we are held in our ordinary waking state through the hyperactivity of a survival mechanism brought about by the fear of death. This hyperactivity shuts out the transcendent consciousness. If we come close to death, the survival mechanism is no longer relevant and relinquishes its grip, so that sometimes, upon being restored to health, the activity of the survival mechanism resumes without being hyperactive. In this way, the transcendent consciousness characteristic of endarkenment could become the ordinary state. However, Wren-Lewis has said that he has not had any advice for realizing the endarkened state and that spiritual strategies used for its cultivation have appeared to have been, in fact, counterproductive (Wren-Lewis, 1994).

Transcendent States of Consciousness

Transcendence, in general, simply refers to a state of being that is in some sense superior to ordinary everyday experience (Baruš, 2003). And indeed, there have been schemata in which various types of transcendent events have been laid along a continuum from least to most transcendent (Cf. Waldron, 1998). However, when I use the expression "transcendent states of consciousness" I am usually referring to mystical experiences as those were defined in Part I, which include a sense of unity, emotional well-being, and noetic value. Wren-Lewis' endarkenment, in which existential questions were purportedly answered, is an example of a persistent transcendent state of consciousness.

It is the noetic quality of transcendent states that poses perhaps the greatest challenge to conventional science, in that inner knowledge about the nature of reality apparently becomes available to the person for whom transcendent states of sufficient profundity have occurred. That knowledge undermines the materialist world view characteristic of scientism. And yet, the controversial thesis of this part of the book is that *science itself can be used as a spiritual practice for seeking transcendent states of consciousness.*

Let us start by examining the ideas of the American philosopher Franklin Wolff. That will lead us into questions about the validity of transcendent knowledge, the psychological dynamics of mathematical activity, and the manner in which mathematics can effect self-transformation. We will conclude with a critical commentary. Although there are references to mathematical constructions in the following text, it is expected that all interested readers should be able to follow the train of reasoning and, hopefully, benefit from the discussion.

Wolff's Enlightenment

Franklin Fowler Wolff, who sometimes identified himself as Franklin Merrell-Wolff by adding his first wife's surname to his name, graduated in 1911 from Stanford University with a degree in mathematics and minors in philosophy and psychology. It was subsequently while he was a graduate student at Harvard University that he became convinced of the probable existence of transcendent states of consciousness that could not be understood through ordinary thinking. After a year of studying philosophy at Harvard University and then a subsequent year of teaching mathematics at Stanford University, Wolff relinquished his academic activities in order to devote himself to the attainment of firsthand knowledge of transcendent states

(Merrell-Wolff, 1994). His efforts reached fruition in 1936 at the age of 49 with the occurrence of a radical shift in consciousness, which he called awakening, recognition, or realization, and whose effects persisted in some respects until his death in 1985 (Merrell-Wolff, 1994; Leonard, 1999; Barušs, 1996). Subsequently, Wolff has stated that the path that he had taken had been that of mathematics, philosophy, and yoga (Merrell-Wolff, 1995a), and has proposed "mathematical yoga" as a spiritual practice particularly suitable for the West (Leonard, 1999). The word "yoga" literally means "yoking" in Sanskrit (Bowker, 1997, p. 1058), and refers to the means for joining oneself to transcendent reality (Barušs, 1996) — with that means, in this case, being mathematics.

To understand Wolff's ideas, let us begin by trying to fathom the transcendent state of consciousness in which he found himself. To begin with, for Wolff, such a state of consciousness is separated from the ordinary state by a discontinuity. No matter how closely we may approach the transcendent through ordinary cognitive efforts, that state remains unrealized. Wolff has used the metaphor of a convergent infinite series to illustrate the point. For example:

$$1 + \frac{1}{2} + \frac{1}{4} + \frac{1}{8} + \cdots$$

is a convergent infinite series whose sum is 2. The terms of the series represent the workings of the mind in ordinary states of consciousness whereas the sum itself, i.e., 2, represents the substance of the transcendent mind. Transcendent states of consciousness cannot be attained no matter how many thoughts within ordinary dualistic consciousness we have, just as the sum of the series cannot be reached through any finitary addition of its terms (Merrell-Wolff, 1995a).

Wolff has said, in fact, that he has "deliberately passed up and down" between the ordinary and transcendent domains "trying to maintain continuity of consciousness" and that "it could not be done" (Merrell-Wolff, 1995a, p. 51). An "inversion of consciousness" (Merrell-Wolff, 1995a, p. 50) occurs at the moment of shifting from one domain to the other so that "one consciousness blacks out and immediately another consciousness takes over." On the finitary side of the barrier he has found himself to be conditioned as usual by space and time whereas on the infinitary side he has been identified with the ground of being. In the course of these transitions, Wolff has also had "a sense of I ascending and descending", to which he "applied the term *escalating self*" and which consisted of "both types of consciousness running concurrently" (Merrell-Wolff, 1995a, pp. 50–51).

A characteristic feature of ordinary consciousness is assumed to be intentionality—intentionality, not in the sense of intending to do something, but intentionality in the sense that the mind is structured in such a way as to be directed toward the objects that occur within one's experience. In other words, there exists a subject for whom there are objects of thought, such as perceptual sensations, images, memories, feelings, self-talk, and so on. Whether or not those objects correspond to anything physically real is irrelevant for the characterization of intentionality. The idea is that there are objects of the mind over and against a subject (Barušs, 1989).

According to Wolff, the distinction between subject and object that is characteristic of ordinary consciousness disappears in the transcendent domain, so that consciousness is structured differently from the way that it usually is. There is no longer any difference between oneself and the universe so that one knows something through identification with that which is known. Wolff has used the metaphor of ocean currents to try to describe the manner in which transcendent thinking functions. Ocean currents can be distinguished from one another yet are simply part of the ocean water. Trying to capture transcendent thinking in dualistic terms is like immersing an open container in the ocean, with the container representing a particular concept and the ocean water the meanings of that concept. The point is that meanings in a transcendent state cannot be exhausted by concepts but necessarily overflow their boundaries (Merrell-Wolff, 1995b).

Because of the radical differences between the ordinary and transcendent states, Wolff has confined the use of the term "experience" to designate experiences in ordinary states of consciousness and has used "imperience" to refer to events occurring within the transcendent state (Leonard, 1999, p. 295; emphasis removed). I have followed Wolff's lead in referring to events that occurred for him in the transcendent domain as "events" or "occurrences" and shall now also use the term "imperiences" for them while reserving the word "experiences" for events that have the usual subject-object structure.

Prior to his awakening, Wolff was prepared for the realization of transcendent consciousness not to have any consequences within his ordinary consciousness. That turned out not to be the case, however, as he noted various experiential effects. To begin with, there was a change in the base of his consciousness. Whereas prior to the recognition his consciousness had been rooted within the ordinary domain, subsequently he found himself transplanted into the transcendent domain, which seemed to him to be both a more real and more natural placement than the previous one. Corresponding to the shift in the base of consciousness was a change of self-identity, in that the self, which he came to regard as a zero point, became all of space without the presence of any subject-object distinctions. There was a sense of depth in that thoughts of great abstraction and universality occurred beyond which there was an "impenetrable Darkness" that he knew to be the "essence of Light" (Merrell-Wolff, 1994, p. 265). As the transcendent self, he found that he was not constrained by space, time, or causality. With this came a sense of freedom, including freedom from guilt. His awakening had a soteriological effect in that the incompleteness of life associated with solitude was replaced with profound contentment. Other affective features of the transcendent state included feelings of serenity in the face of environmental disturbances, an intense joy that was in the nature of a "force-field" giving a "glow to life" (Merrell-Wolff, 1994, p. 269), and benevolence interested in seeing good brought about in the world. The role of information changed, in that, whereas previously information had been acquired to seek reality, now its purpose was to express the reality that had been found. Indeed, we have already alluded to a different type of knowledge that became activated, and we will consider the nature of that knowledge in greater detail in a while (Merrell-Wolff, 1994).

Associated with the force-field of joy was the presence of what Wolff called the "Current", a fluid force of life, which manifested itself as physiological enjoyment and psychological bliss. Wolff has said that others have reported feelings of joy in his presence even though he had not drawn attention at the time to his own inner state of being (Merrell-Wolff, 1994, p. 19). He has also said that he has asked others to give written descriptions of their experiences while the current has been present, and the states of consciousness that they have described have included features of mystical experiences (Merrell-Wolff, 1970). Doroethy Leonard, Wolff's granddaughter, has said that during a public lecture given by Wolff, she entered a peaceful and profoundly still state. Following the talk she found it difficult to contain the joy that she was feeling. When she drew Wolff's attention to her state of being, he told her that she had been in the current (Leonard, 1995). It appears as though the presence of the current in Wolff could induce transcendent states of consciousness in those who came in contact with him.

Wolff has said that it was easy for him to turn the current on and off, and move at will through the discontinuity between the ordinary and transcendent domains. "There's just a little valve [somewhere] in one's total psyche that I call the butterfly valve. You flip it as easily as you would move a finger..." (Merrell-Wolff, 1970). However, the force associated with the transcendent state created a strain on the physical body. Wolff used the analogy of trying to send 100 amperes of electrical current at high electrical potential through a 10 ampere fuse. There is pressure to use more than the 10 amperes, but if the fuse were to burn out, then the connection with the ordinary world of expression would be broken (Merrell-Wolff, 1994). Indeed, Wolff did not induce the current as frequently during the last years of his life.[13]

After a day of exceptional mental clarity and acuity, a further expansion of consciousness occurred for Wolff in 1936, which he called the "high indifference". This was a state of consciousness in which there was an equilibrium between all dualities, including the duality between ordinary consciousness and the awakening of his initial transcendent consciousness as described earlier (Merrell-Wolff, 1975). Wolff's thinking remained alert for the several hours during which this state lasted, and he felt himself identified both with a primordial "unlimited and abstract Space" as well as a "subject-object and self-analyzing consciousness" that had a "sort of point-presence within that Space" (Merrell-Wolff, 1994, p. 284).

For Wolff, as for Wren-Lewis, there was a fundamental shift in consciousness away from individual identity toward a ground of consciousness that gives rise to the manifested world. Wolff and Wren-Lewis, when identified with the universal substrate, felt that they knew the objects of the world through having become one with them. They experienced exceptional emotional well-being. And both could re-enter the transcendent state by an adjustment of consciousness. Wolff's enlightenment and Wren-Lewis' endarkenment both speak to the possibility of the existence of ongoing, rather than just transient, transcendent states of consciousness.

Transcendent Knowledge

How do those for whom transcendent states of consciousness occur know what is happening in those states? Well, this takes us back to our discussion about inner knowledge from Part II. According to Wolff, we ordinarily consider the activity of two sources of knowledge: sensory perception and ordinary thinking. As we have already mentioned, for him there is also a third way in which knowledge is possible, namely, knowledge through identity, which he called "introception" (Merrell-Wolff, 1995b, p. 143). Let us further characterize the nature of this transcendent knowledge.

Introception gives certainty of that which is known although that certainty cannot be conveyed through conceptual thought to anyone else (Merrell-Wolff, 1995b). The effect of introception on the "personal mind is that of unequivocal demonstration not unlike nor less convincing than rigorous mathematical demonstration" (Merrell-Wolff, 1995b, p. 277). According to Wolff, there is logical inevitability to the products of higher thinking. At its own level, introception has more the character of seeing than of conception. However, for Wolff, there is nothing perceptual about this "seeing". Rather he says that this character of seeing has more to do with the "sense of seeing an idea" (Merrell-Wolff, 1995b, p. 154).

Wolff's notion of seeing reminds me of what happened in an undergraduate course in relativity theory that I took. The professor would write an equation on the chalkboard and then turn to us, the students in the class, and simply remain silent. After a while, he would point to the equation and ask "Do you see that?" What he meant, of course, was whether we understood what the equation revealed about the nature of reality.

The relationship of physical sight to this notion of seeing was brought out in a retrospective study of sight during out-of-body and near-death experiences of the blind. The first, perhaps rather surprising finding was that some of those who were blind could see during such experiences. This included those who had been blind from birth and whose brains would not have developed the capacity for sight. Some of the experiencers could not definitively identify that what they had experienced had been sight, whereas others could. However, upon closer inspection of the data, the researchers found that there were differences between ordinary sight and the sight of the blind, in that sight during the anomalous experiences had more the character of simply knowing, rather than seeing. The authors of the study used the term "transcendental awareness" to refer to this hybrid of seeing and knowing (Ring & Cooper, 1997; quotation from p. 140, emphases removed).

Introception can influence a person without her awareness (Merrell-Wolff, 1995b) but becomes explicitly activated with the occurrence of transcendent states of consciousness. How can this explicit activation occur? According to Wolff, in our ordinary dualistic state, consciousness has the quality of a vector in that it flows outward from an inferred subject toward an object. Ordinarily, we conceptualize the flow of information as being in the opposite direction: inward, from the objective world, through sensory mechanisms, into the brain. For Wolff, there is an outward flow of consciousness that can be reversed "by turning it through the angle π, so that it returns toward its source without projecting an object in consciousness, no matter how subtle" (Wolff, quoted in Leonard, 1999, p. 49). When Wolff has used the expression "through the angle π", he has been referring to an angle of π radians, which is the same angle as 180 degrees. The outward flow of consciousness is metaphorically turned through π radians in such a way as to allow it to rest in the pure subject. The subject is realized, not through apprehending the subject as an object of thought, but through identification with that which is known. The introceptive faculty arises with such reversal of the direction of consciousness (Leonard, 1999; Merrell-Wolff, 1995b).

I find it easy to explicate Wolff's descriptions of the reorientation of consciousness toward its source, although Wolff has said that it was difficult for him to consciously recognize and conceptualize the process (Merrell-Wolff, 1995b). But it is one thing to say that consciousness can be turned toward its source and quite another to actually do so. One way of trying to get a grip on that to which Wolff is pointing is to consider the sense of existence that one can have for oneself, which accompanies the contents of our experiential stream. Of course, thoughts that we have about our sense of existence are just more objects for consciousness. But even as we think about our sense of existence, we can note the felt aspect of existence that is part of our ongoing awareness.

Noting that feeling could again simply become part of the experiential stream but there is something to which it actually corresponds, namely an actual sense of existence that occurs for us. I think that it is the non-objective sense of existence to which Wolff is pointing. Of course, the reader may find my way of understanding the reorientation of consciousness as opaque as Wolff's. Needless to say, ultimately, each person is left with her own way of understanding this reorientation on the basis of her own consciousness.

Although there is a discontinuity between the ordinary and transcendent states, the discontinuity itself can be progressively approached from the side of ordinary dualistic consciousness. Wolff has claimed that a key factor in his awakening was the realization that ontological fullness was present precisely to the extent that sensory and conceptual accessibility were missing. In other words, the less we are able to grasp something in a concrete manner, the closer it is to reality. For Wolff, the presence of that which is ontologically substantial is inversely proportional to that which is ponderable. Or, using an analogous expression, reality is inversely proportional to appearance. Substituting their initial letters for "reality" and "appearance" we get the hyperbolic function:

$$R = \frac{1}{A}$$

For Wolff, in a reversal of the usual materialist supposition, the real lies in the direction of abstraction, not concretion. Concepts that are more abstract, such as mathematical constructions, are closer to the real than more concrete concepts or the sensations of perceptual experience (Merrell-Wolff, 1995a).

What does it mean to say that there is a discontinuity if there is a gradual approach to that which is real? A metaphor from my days of working as a roofer comes to mind. I would climb a ladder to get up on a roof. I found that there was a point at which I deliberately transferred weight-bearing responsibility from the ladder to the roof. As I climbed and even as I reached out with one leg for the roof, it was the ladder that I relied upon to bear my weight. At some point, I would have established enough traction on a roof with the one leg that I placed on it, in order to be able to transfer the weight-bearing responsibility to the roof, even though my other leg was still on the ladder. I recall explicitly noting the point at which the weight-bearing responsibility would shift. That point represents a discontinuity. Climbing the ladder is a continuous process reflective of incremental increases in altitude, but the climber is still dependent on the ladder. Once weight has been shifted to the roof, it is the roof upon which a person depends for bearing her weight.

Transcriptive Thinking

As I have already said, for Wolff, transcendent knowledge has the character of water in ocean streams rather than water in containers. But out of this ocean of thought arises another type of thinking that is a blend of introception and conception, in that it consists of abstract thoughts of the ordinary domain yet nonetheless containing several "drops of the supernal waters" (Merrell-Wolff, 1995b, p. 277). Introception can find some embodiment in conception in the form of "transcriptive" thinking (Leonard, 1999, p. 257). In other words, transcriptive thinking refers to "the process of thinking and resultant formulations of thought in which conceptions are containers for introception" (Barušs, 1996, p. 161). The primary example of transcriptive thinking relevant for our discussion is that of mathematical reasoning.

Wolff has used the notions of "thick" and "thin" to elucidate the semantic nature of concepts in the context of the inverse relationship between substantiality and ponderability. Concepts with large values of A in the hyperbolic function could be designated as being perceptually thick but introceptually thin. Examples would be concepts such as sticks, stones, brains, and hockey pucks. They have as their meanings references to particular perceptual experiences. That is to say, such concepts point to something other than themselves and have little of actual substance to them.

In contrast to this are concepts with small positive values of A, which could be regarded as being perceptually thin but introceptually thick. Examples would be beauty, sets, and topoi. As A goes toward zero, the significance of these concepts lies less in pointing to objects and more within the concepts themselves. That is to say, the concepts inherently enrobe their meanings. As the limit of perceptual thinness is approached, thought passes beyond verbal and symbolic formulation and becomes "disembodied Meaning" (Merrell-Wolff, 1995b, p. 170). Such thought is introceptually virtually infinitely thick.

According to this account, we miss the real significance of thinking if we disregard the introceptual component of a thought and pay attention only to its perceptual aspect. For example, if, on the one hand, we regard consciousness as information in an information-processing system, the meaning of the concept "consciousness" is directed toward something that can be grasped perceptually and consciousness becomes just another physical event. On the other hand, if we conceptualize consciousness as the sense of existence of the subject of mental acts, then meaning is directed back toward the mystery of existence within our experiential landscape.

For some readers, the example of consciousness in the previous paragraph may appear to be backwards, because information may seem to be a highly abstract concept, whereas the feeling of existence can be regarded as a perceptual experience. However, the concept of information usually has a concrete referent in that it has been used to designate distinctions between physical settings of a computational device. In contrast, the sense of existence of the subject of mental acts, for all that it sounds like a perceptual experience, has more the character of an abstract reality. The reader can determine for herself which of these versions of consciousness has the greater introceptual quality for her (cf. Barušs, 1987; 1990a). Incidentally, for Wolff, consciousness is the immaterial substantial substrate underlying all of experience (Merrell-Wolff, 1995b).

The movement toward greater reality associated with the increased introceptual content of transcriptive thinking not only allows for increased inherent meaningfulness but also for increased depth. For Wolff, depth cannot be conceptualized or identified as part of experience, but can be felt in genuine religious experiences and directly realized by introception. Depth is the inexpressible element of transcendent states of consciousness that makes them what they are. Any effort to express depth fails to convey its essential character and can always be interpreted in such a way that the very characteristics one aims to convey are omitted, giving the impression that there is only that which lies on the surface.

This last is an important point. A number of times in this book I have alluded to the notion of "understanding" in such a way as to try to preserve its depth of meaning. From a contemporary computational approach to the mind, my notion of understanding would itself be reified as specific information processing in the same way that consciousness has been reduced to tangible events (cf. Dennett, 1978; 1988). This is not to say that there is no informational component to understanding. That is not the point. The point is that neither meaning nor depth reside in information but both meaning and depth are possible for someone for whom understanding occurs. For anyone who insists on remaining on the surface, this last sentence can again be reified in information-processing terms denying any ontological independence to meaning or depth. If Wolff is right, then as long as one keeps reifying in information-processing terms concepts whose purpose is to characterize something deeper, then transcendent states of consciousness will necessarily elude her.

Mathematical Thinking

We can read all the accounts about a far away land that we can find, but how are we to know that the explorers did not end up in a house of mirrors instead of the ground of being? In the end, there is only one way to know, and that is to do our own research to see for ourselves. Wolff has said that proof is possible only through direct realization of the transcendent reality (Merrell-Wolff, 1995b). And indeed, as I have already mentioned, he has proposed mathematical yoga as a particularly suitable manner of seeking transcendence for those who identify with Western culture. So let us turn then to an explication of mathematical yoga.

Wolff has said that "mathematic is that portion of ultimate truth which descended from the upper hemisphere... into the Ādhāra with minimum distortion and thus become the Ariadne thread by which we may ascend again, most directly, most freely".[14] The word "Ādhāra" is the Sanskrit word for "container" and refers to the manner in which the transcendent self is embodied in its sheaths according to Hinduism (Bowker, 1997, p. 17). In other words, mathematics is the purest expression of the transcendent within the ordinary dualistic consciousness and can be used as the most direct means of access to the transcendent. I find Wolff's use of the singular "mathematic" rather than the plural "mathematics" interesting, in that it suggests a coherence to mathematical constructions that we do not usually associate with them. It is as though there were an essence of mathematics giving rise to its multifarious manifestations.

I also find Wolff's image of an upper hemisphere to be interesting. At one point when I was studying mathematics and before I had come across this passage from Wolff, a spontaneous impression arose in my mind. In my imagination I had climbed to the top of a vertical ladder that terminated at the bottom of an enormous hemisphere. I had poked my head and shoulders through a round hole so that I was looking up through the hollow interior of the hemisphere out into infinity. I stood in awe. It was the world of mathematics, complex and vast, stretching above me without limit. I was humbled by the insignificance of any effort I could ever make to really know it.

I have found that occasionally there is a misperception on the part of some people concerning the nature of mathematics. Mathematics is sometimes identified with arithmetic calculations, equations, or computer algorithms, often arousing feelings of tedium. But such productions are just a visible mask of mathematics. In my experience mathematics consists of precisely delineated concepts and the relationships between them. It was not until I had progressed a ways in my study of mathematics that I was able to see that. In the imaginary impression that I just related, I had already climbed a considerable distance vertically just to get to the bottom of the hemisphere representing the world of mathematics. The more accessible visage of mathematics with which it is sometimes identified is, of course, also a part of mathematics, but gives a misleading impression of its nature. In my imagination, I had found the lower hemisphere—the lower boundary of mathematics. Wolff's symbolism suggests that there is an upper hemisphere from which mathematics emerges thereby symbolically completing the sphere.

Wolff has not been the only mathematician to suggest a role for mathematics in spiritual aspiration. Hermann Weyl has said that "... purely mathematical inquiry in itself ... by its special character, its certainty and stringency, lifts the human mind into closer proximity with the divine than is attainable through any other medium" (Weyl, 1932/1989, p. 7). Charles Musès has gone even further to say that it is "comparatively easy" to produce "intensely exhilarated states of consciousness ... by means of mathematical ideas" although it is not clear through what mechanism that is to occur (Musès, 1970; quotations from p. 46). How is it that mathematics could provide such direct access to transcendent states of consciousness?

Following ideas suggested by F.S.C. Northrop (Northrop, 1946/1966), Wolff has maintained that there are continua, specific to the East and the West, along which an aspirant can move from the "differentiated" relative domain to an "indeterminate" transcendent domain. Spiritual aspirants in the West have often adopted the methods of the East and tried to move along the continuum belonging to that culture rather than recognizing the possibility of exploiting the capacity for theoretical thinking that has been developed in the West. According to Wolff, aspirants in the West could effect a transition to transcendence by moving along a theoretic continuum from its differentiated pole, represented by science and mathematics, to an indeterminate, transcendent pole. In other words, self-transformation not only does not require setting aside what one does as a scientist, but of using those activities as a basis for further development toward the realization of transcendent states of consciousness (Merrell-Wolff, 1995a; quotations from pp. 46 & 47).

Perhaps the reader can see where this discussion is heading. The theoretic continuum is just the inverse relationship between substantiality and ponderability, with the indeterminate pole of the theoretic continuum coinciding with the value of ponderability set at zero and the value of substantiality reaching infinity. Theoretical concepts, such as mathematical constructions, are so abstract that there is nothing left for the senses at which to grasp. Hence they are close to the type of knowledge that exists in transcendent states of consciousness. If one can rarefy one's thinking sufficiently, one's thinking is no longer "completely definable" (Merrell-Wolff, 1995a, p. 35) but becomes "transcriptive" (Leonard, 1999, p. 257). According to Wolff, such thinking has both effortful and autonomous components, so that it is a blend of transcendent content and relative form (Merrell-Wolff, 1995b).

But is transcriptive thinking the way that mathematicians actually think? On the one hand, George Lakoff and Rafael Núñez have proposed a "cognitive science of mathematics ... to characterize precisely what *mathematical ideas* are" and have proceeded to demonstrate how basic schemata inherent in everyday interactions with the physical world could be compounded through the layering of metaphors to produce mathematical abstractions all the way to a special case of Euler's formula:

$$e^{\pi i} = -1$$

According to Lakoff and Núñez, essences, including those of mathematics, are just products of human minds (Lakoff & Núñez, 1997; 2000; quotation is from 1997, p. 21). On the other hand, Peter Dodwell has argued that "cognitive science has developed far too restricted an image of humankind, a truly pusillanimous concept of mind" incapable of adequately explaining mathematical thinking, and has supported the notion of a world of ideal forms (Dodwell, 2000, p. 176). Implicit cognition of the sort hypothesized by Lakoff and Núñez probably plays a role in mathematical thinking, although Euler's formula is still a long way from the abstractions of contemporary mathematics. Dodwell has done well to point out the significance of the meaning of mathematical constructions which is not captured by the kinds of cognitive processes postulated in contemporary cognitive science. In terms of the material-transcendent dimension from Part I, whereas Lakoff and Núñez represent a materialist approach to mathematics, Dodwell's approach is transcendent. Given that no one has empirically studied how mathematicians think, the question of what they actually do remains open.

Mathematical thinking has both implicit and explicit components but it is not clear to me what those are. In my experience, doing mathematics apparently entails more than the usual processes of cognition associated with everyday consciousness. There can be images as well as self-talk present, but these are inadequate, in themselves, for understanding mathematical constructions. The experience of doing mathematics for me has been one of directly manipulating tenuous concepts without any apparent perceptual, verbal, or metaphorical mediation. Working with mathematical constructions is both abstract and precise. Once a train of mathematical thinking has been completed, it can take considerable effort to put it into a form that can be communicated to others. Wolff's contention that mathematical thinking of sufficient abstraction is a type of transcriptive thinking makes sense to me on the basis of my own experience.

Wolff has said that we are locked into a "participation in the object ... that is harder to break than bars of iron". A sustained effort of the will is necessary to overcome this obstacle to realization of transcendent states of consciousness.

However, the austerity required to do mathematics is the same as that which can awaken the introceptive function. And much of the labour required to break the hold that objects have over the mind has already been done by the person who has learned to work with concepts of such tenuity as those found in mathematics. Thus, the proximity of transcendent states has been increased (Merrell-Wolff, 1995b; quotation from p. 166).

Mathematical Yoga

According to Wolff, the contents of everyday consciousness turn out to be a matter of appearance. However, that which is real can be known in transcendent states of consciousness through introception. And there is a continuum from the apparent to the real that can be traversed using our capacity for theoretical thinking. This is the basis of mathematical yoga, the mathematical means of realizing transcendent states of consciousness.

However, Wolff had experienced states of consciousness with extraordinary value through his involvement with mathematics but not enlightenment (Leonard, 1999). Mathematical yoga involves more than just doing mathematics with a high degree of abstraction. There are two other necessary ingredients: philosophy and yoga. Wolff has included philosophy because, according to him, philosophy is that discipline which is concerned with establishing meaning and value. In doing mathematics with spiritual intent, one cannot proceed mechanically, but must consider the meanings inherent in the mathematical constructions. In effect, the point is to establish the depth of meaning as we have already discussed. Or to put it another way, mathematics is to be done in a way that authentically develops one's understanding (Merrell-Wolff, 1995a).

For the mathematical path to be effective in uniting a person with the transcendent, according to Wolff, there must also be a change of attitude from "self-withholding" to "self-giving" (Merrell-Wolff, 1995b, p. 150). Indeed, through proper practice of mathematics and philosophy, one can reach the door to the transcendent but cannot force it open. *The transcendent is not compelled by our actions within the dualistic realm of consciousness but is itself the causal agent.* However, one can solicit the transcendent by a "complete sacrifice of everything that [one] is and has" (Merrell-Wolff, 1995a, p. 33). It is this to which Wolff refers in saying that yoga is a necessary ingredient of the mathematical path.

As outlined above, much of the work of liberation from the confines of ordinary consciousness has been done by developing the capacity for abstract thinking, such as that required of mathematics, and of doing such thinking with a view to understanding the subject matter. But whereas the intellect may have been loosed from its hold on the objective world, the same may not be true of the will and feelings. From the point of view of an aspirant, the required movement of consciousness away from the ponderable universe can appear to be self-annihilation. It is in this context that, according to Wolff, it is necessary to sacrifice everything that one is in order to relinquish one's hold on the objective surface of consciousness. In spite of his occasional use of behavioural terminology, it seems that, for Wolff, such sacrifice is an attitudinal shift rather than a behavioural act so that, having made the sacrifice, everything beneficial is returned with the difference that one is now a steward rather than owner of private possessions (Merrell-Wolff, 1995a; 1995b).

This mention of sacrifice may seem somewhat out of place in science, but it is a quality with which I imagine most scientists are quite familiar. In addition to the incidental sacrifices of time, money, family life, and sometimes one's health and sanity, there is a fundamental role that sacrifice plays for a scientist if she is to be authentic. In the course of any scientific study, the investigator must give up any expectations that she may have about the outcome of the study, and accept whatever the results happen to be. She must sacrifice herself to reality the way that it actually is and not the way that she has been led to believe that it is or the way that she would like it to be. Her motives must be pure. In other words, it seems to me that, for a scientist, the sacrifice is to the truth. In particular, in the case of spiritual aspiration, it would mean giving up whatever preconceptions one might have about the nature of transcendence and accepting whatever might be revealed about it in a transcendent state of consciousness.

I think that there is another way of understanding the process of sacrifice to which Wolff has alluded. For Roberto Assagioli, one of the keys to self-transformation was the capacity for disidentification. The idea is to disidentify with the personality elements and to identify with the self. In fact, ultimately the identification is to be with the higher self. Assagioli has suggested an exercise of progressive disidentification: "I *have* a body, but *I am not* my body. . . . I *have* an emotional life, but *I am not* my emotions or my feelings. . . . I *have* an intellect, but *I am not* that intellect". Then one is to identify with the invariant: "I am I, *a centre of pure consciousness*" (Assagioli, 1965, pp. 116–117). It seems to me that this would invoke the same psychological dynamic as the self-annihilation of sacrifice proposed by Wolff.

In describing mathematical yoga I have at times referred specifically to mathematics and at other times more generally to science. Is mathematical yoga just for mathematicians? This is a question that colleagues interested in mathematical yoga and I have deliberated extensively. We think that mathematical yoga can be practised more generally by scientists and those with a scientific perspective. For Wolff, the point was that in the Western culture we have developed a theoretical approach to the world that can be exploited for the purposes of spiritual aspiration. It is the capacity for abstract thinking which is the key to mathematical yoga and not the mathematics itself. Having said that, it could also be that there are aspects of mathematical activity that, when present, make it uniquely effective.

So, now we have a recipe for scientific self-transformation. We think meaningfully in such a way as to get away from the ponderable toward the substantial along the theoretic continuum into the realm of transcriptive thinking while sustaining an attitude of absolute sacrifice. We do so until such time as a shift toward transcendent consciousness occurs. Does the recipe work?

The Agency of Invariants

In considering the practical utility of mathematical yoga, the place to start is with Franklin Wolff himself. For Wolff, the realization of the inverse relationship between substantiality and ponderability, which he expressed in mathematical form, was a decisive factor in his awakening. However, it did not result in enlightenment. He has said that it took him a few days of effort to clear up the "residual barriers to the complete identification of the self with the supersensible and substantial world" (Merrell-Wolff, 1994, p. 262). This he did by metaphorically turning consciousness back through π radians toward its source. In other words, in addition to whatever ingredients of mathematical yoga he had previously employed, in the end, Wolff turned to a meditation technique that depended upon a reorientation of consciousness to the subject of mental acts.

How are we to understand this activity? Is mathematical yoga incomplete? Is the reorientation of consciousness something that we need to do as well? The problem is that we are unlikely to succeed at reversing the flow of consciousness to find the self as the pure subject. According to Wolff, we are psychologically too "objective" for the self to be readily accessed. So what are we to do? It may be instructive to note that the self functions as an invariant with respect to objects in the quest for the transcendent so perhaps it is enough to use any invariant. Wolff has said that an invariant is needed within the flux of ordinary experience lest we become lost. The discussion here is not about getting lost but about precipitating a shift in consciousness, but perhaps the same agency of invariance applies. At any rate, for Wolff, the role of the invariant for the Western mind is played by pure mathematics (Merrell-Wolff, 1994; quotation from p. 351).

The problem with considering mathematics as an invariant is that mathematics was deconstructed during the twentieth century. In 1931 Kurt Gödel created first-order statements in the language of arithmetic that were true but could not be proved from elementary axiom schemata for arithmetic. Something similar happened to set theory, when Gödel found in 1938 that the absence of any sets of cardinality between $\aleph_0$ and $\aleph_1$ was consistent with the axioms, and Paul Cohen found in 1963 that the existence of a set of cardinality between $\aleph_0$ and $\aleph_1$ was consistent with the axioms of set theory (D. Nelson, 2003; cf. Huber-Dyson, 1991; Bell, 1977). Which is it then? Are there such sets or are there not?

Although not all mathematicians would agree (Huber-Dyson, 1991), the net result of this deconstruction has been a sharpened awareness that there is no single correct mathematics but that we can create the mathematics that we want through selectively choosing some axioms rather than others. Wolff was aware of this argument. His contention was that, having chosen one's axioms, one is necessarily constrained by the laws that govern the deduction of theorems from the axioms. Here was something to which we could cling. Or at least, to which we could cling until we realized the transcendent state of consciousness at which time such invariants themselves would disappear (Merrell-Wolff, 1994).

But there is also a problem with reliance on logic. Not only was mathematics deconstructed during the twentieth century, but so was logic. We usually accept a standard set of rules for how to form conjunctions, how to quantify over variables, how to make inferences, and so on, which is known as classical logic. The formalization of such procedures is the classical predicate calculus.

However, there is nothing sacrosanct about classical predicate calculus. In particular, what about the law of the excluded middle which states that for any well-formed formula χ of a first-order formal language the disjunction χ or $\neg \chi$ can be used in the process of reasoning? For example, if χ is the proposition that "the cat is black" then we could argue that "the cat is black or the cat is not black". If we refuse to accept this logical axiom schema but retain the remainder of a suitable classical axiomatization, then we end up with intuitionistic predicate calculus (Rasiowa, 1974; Heyting, 1956/1976), which, it has been argued, is the "logic of scientific research" (Grzegorczyk, 1964, p. 596).

Outside the context of mathematics, there are more radical ways in which logic can break down. We can quibble about the law of the excluded middle for, after all, how much of a difference does its exclusion really make to the substance of mathematics. But what if we were to reject non-contradiction and allow contradictions of the form χ and $\neg\chi$ (cf. Huber-Dyson, 1991)? Our feline example would become "the cat is black and the cat is not black". Indeed, if we were to derive a contradiction then we would usually assume that the premises from which we started the argument must have been incorrect. Let us consider some examples of situations in which classical logic breaks down.

Benny Shanon conducted interviews with 178 informants who had ingested the psychedelic ayahuasca and he himself had consumed the brew more than 130 times over the course of 10 years. Drinkers typically reported having encountered an apparently objective, non-ordinary reality leading them to a metaphysical view that Shanon has characterized as "idealistic monism with pantheistic overtones" (Shanon, 2002, p. 163). Ayahuasca, according to Shanon, has its own rules of non-ordinary logic by which the ayahuasca experience can be described. In particular, the middle is not excluded in Shanon's logical law of "disjunction" for which he has stated that "between p and *not p* there is always a third possibility", and contradiction is permitted in his law of "conjunction" for which he has stated that "both p and *not p* are true". For Shanon, "in an ultimate sense, truth is the conjunction of statements that are normally regarded as contrary to one another" (Shanon, 2002, p. 169).

Wolff has made a point similar to Shanon's rule of disjunction. According to Wolff, dichotomies often arise in Buddhist teachings that are resolved, not by choosing one or the other of the alternatives, but by choosing neither. In other words, for a statement A, neither A nor not A is true. For Wolff, in particular, it is the subject of experience which cannot be defined, named, or classified into dichotomies (Merrell-Wolff, 1995a). And Wolff has also explicitly stated that "the logical principle of contradiction simply had no relevancy" during his imperience of the high indifference. As soon as he would try to isolate anything using his self-conscious thinking it would become transformed into its opposite (Merrell-Wolff, 1994, p. 286).

Both Shanon's and Wolff's examples of the breakdown of logic apply to a transcendent domain, so the question remains of whether there is anything invariant about logic within dualistic consciousness. I think there are two answers to that question. The first is that there could actually be a preferred logic, perhaps along the lines of intuitionistic predicate calculus, which could reliably guide us to the transcendent. The second answer is that it may not matter which rules of logic we follow; it is the following of rules which is invariant. And there may be some merit to both of these ideas. But is there not yet something more reliable to which we can cling?

In some ways mathematics is not as arbitrary as I have made it appear to be. Wolff has hinted at the notion that some numbers, such as π or e, are substantial in a transcendent sense (Leonard, 1999). The real number π is, of course, the ratio of the length of a half-circle to its radius. Clearly this is an idea that we can have in our minds and so it is a thought among other thoughts that we can have. However, in thinking about this ratio, we can realize that it cannot be anything other than what it is, as long as we confine ourselves to the Euclidean plane. We cannot change that ratio by creatively imagining it to be something else. In that sense, it is objective. What is it about reality that results in a specific fixed ratio of the length of a half-circle to its radius?

But this line of thinking brings us back to the more general question of the ontological status of invariants such as π. Materialists, for whom such mathematical ideas are simply the by-products of physical processes of an objectively real brain, keep pushing the discourse to greater levels of complexity by suggesting ways in which such notions could be brain processes. There are no perfect circles in the world other than through abstraction from sensory experiences. Transcendentalists, on the other hand, gather the discourse back up again and argue that the materialists' elaborate explanations lack substance. Why does the bowl have a circumference in the first place for it to give rise to the idea of a circle? For that matter, why is there anything rather than nothing? For Wolff, contrary to a materialist position, it is meaningless to posit the existence of reality independent of consciousness, "since no meaning attaches to the notion of existence apart from consciousness" (Merrell-Wolff, 1995b, p. 184). Meaning is established in consciousness and it is there, within dualistic consciousness, that the value of π exists as objectively real.

But the failure to agree upon the ontological status of π does not detract from the functional significance that π as an invariant could have for our efforts at self-transformation. The problem is that despite its invariant nature, it is not clear how π can help us to realize transcendent states of consciousness for all that it is, ironically, a "transcendental number."[15] Is it possible that, rather than metaphorically turning consciousness through π radians to find the invariant self, it is enough to find the invariant π? And then do what?

A Scientific Resolution

It seems to me that mathematical reasoning naturally involves concentrative meditation in that, in doing mathematics, one's mind rests on the mathematical constructions about which one is thinking. I found in my experience, that I could sustain attention without interruption on a mathematical problem for hours at a time. The analogy that I have used for this is that of a pin set on its point without falling over. And for me such thinking proceeded in an abstract yet precise manner that is not adequately characterized by our usual notions of thinking in the ordinary waking state.

Furthermore, concentrative meditation has been associated with transcendent states of consciousness. As one persists in seeking to understand the object of one's attention, one's understanding could be infused with insights that could be regarded as originating from a transcendent dimension. Or, as is sometimes alleged, one could be drawn into the transcendent dimension through identification with that which is known (cf. Baker, 1975). In other words, doing what we do as scientists by contemplating abstract ideas using some variation on logical reasoning could lead to transcendent states of consciousness.

Why, then, are we not all enlightened? Well, as we have already considered, all three aspects of mathematics, philosophy, and yoga must be engaged. But there is yet one more possibly critical ingredient, which is related to Wolff's notion of yoga. Perhaps part of the answer lies in the need for intensity of purpose. For Wolff, the study of one's subject matter must be pursued with complete investment of oneself in order for it to have transformative effects (Merrell-Wolff, 1995b). In fact, the effort to seek that which lies underneath the surface of life may be the only really necessary ingredient. When asked for the keynote leading to his success, Wolff's response was that "it might've been the tendency to drive toward the root . . . that from which all comes" (Wolff, interviewed by Bray, 1980). Of course, even that may not be strictly necessary in some cases given that transcendent states of consciousness can emerge apparently unsought as evidenced by the experiences of Wren-Lewis.

So, what do we need practically in order for enlightenment to occur? How minimal can science be in order to function as a spiritual practice? Can transcendent states of consciousness eventuate just as a consequence of the open-minded development of understanding in an effort to drive toward the root? Can such an incremental process trigger the inversion of consciousness apparently necessary for transcendent states of consciousness to occur? Can such perseverance in one's quest lead to a point of radical transformation characterized by awakening?

There is an interesting treatise about meditation couched in arcane ideology written by Alice Bailey. My interpretation of the substance of her argument is that academic activity is a form of meditation in which the gradual expansion of understanding eventually leads to a radical transformation of identity and the occurrence of transcendent states of consciousness (Bailey, 1922/1950). So we have at least one attestation that the pursuit of understanding in itself can lead to enlightenment.

But, of course, from a scientific point of view, recipes and attestations are merely hypotheses to be empirically tested. We can try to identify the necessary ingredients of scientific self-transformation through experimentation. This experimentation can be done using conventional forms of investigation or, as discussed in Part I, carried out by a researcher for herself. And there can be hybrids of personal and conventional methodologies, one of which I will briefly describe here.

Charles Tart has proposed "state-specific sciences" whereby researchers would individually enter the altered state of consciousness that is of interest to them and make observations, theorize, predict, and communicate within that altered state. In doing so, the characteristics of individual researchers would be noted so as to compensate for biases in making observations. Alternate logics that are appropriate for the altered state would be employed in theory construction. Based on the theories, predictions about new events would be made and theories revised on the basis of the observations made. Full communication between researchers in altered states would be necessary but could take forms that appear to be nonsensical to those outside the altered state. In this way, a consensual understanding of the nature of reality in altered states of consciousness could be developed. As an example, Tart has pointed out that mathematics already functions as a state-specific science (Tart, 1972; 2000). In the same way, individual scientists can deliberately seek transcendence, work together with one another, and then report the results of such investigations as best they can to the rest of the scientific community so that we can know more about the nature of consciousness and reality.

A Critique of Enlightenment

Over the years, as I have taught about transcendent states of consciousness in my classes, students have asked what the point is of entering such states; why anyone would want them to occur. Are they even a good thing? Having such spiritual experiences does not appear to necessarily make someone into a better person in some reasonable sense of "better". And what about the sense of unity with the universe that is characteristic of mystical experiences? Do we really want to live in a world in which all selves are ultimately a single self? "Is that heaven or a solipsistic hell?" (Horgan, 2003, p. 230). And exactly just what knowledge does enlightenment convey? Is it valid?

Let me start with the problem of regarding those who have experienced transcendent states of consciousness as freaks. From the perspective of those for whom these experiences have occurred, they do not feel somehow as though they have been removed from humanity, but rather, as having been returned to it. The Christian contemplative, Bernadette Roberts, after years of arduous transformation, has said that she realized that we were intended to live our productive years and to undertake our responsibilities in a state of unity with God (Roberts, 1991; 1993). For Wolff, the person who has realized the nature of reality is free to participate in the world of objects or that without objects (Merrell-Wolff, 1994). In fact, he has made the point that such people could well be in our midst without being recognized (Merrell-Wolff, 1995a). The contention, on the part of those for whom they have occurred, has been that transcendent states of consciousness are not the end of the journey but the beginning of authentic humanity.

In addition to the emotional benefits of awakening as previously described, there is also a personal aspect to transcendent states of consciousness that could be a source of satisfaction. We are often comfortable thinking of ultimate reality as having an impersonal character. Such impersonality comes through in the accounts of both Wolff and Wren-Lewis. But there could also be a personal aspect to ultimate reality. For Wolff, just as we are physically interpenetrated by space, so our being is interpenetrated by an underlying divinity. However, unlike the coldness of space, such divinity is characterized by warmth. And, upon introceptive realization, that divinity "is found to be quite the most intimate of all things" thereby contributing to the soteriological value of realizing its nature (Merrell-Wolff, 1995b, pp. 210–211).

Perhaps the most compelling reason for seeking enlightenment is the value associated with knowledge. Wolff has referred to a notion that he has attributed to Śaṅkara, namely that "when one realizes a given condition is an illusion, it not only ceases to be, but ceases ever to have been". Wolff has used the analogy of a mirage to illustrate this. Suppose, he has said, that one were to see a beautiful lake as one is walking in the desert. As one continues to walk, one eventually realizes that there is no lake and that the image was a mirage (Merrell-Wolff, 1995a; quotation from p. 23). Similarly, upon awakening to the transcendent state of consciousness, one apparently realizes that the world is not as it has always seemed — that one has been misperceiving the nature of reality all along. At that point it is not that the world as ordinarily understood has ceased to exist, but rather it ceases to have ever existed as such. For someone who is interested in knowledge, understanding the truth of the matter is important.

A noetic quality is present in transcendent experiences, drug-induced states, alien abduction experiences, and other alterations of consciousness. But is it really noesis? Conviction, in and of itself, is not a good criterion of truth. There appears to be too much conviction and not enough actual knowledge around. People tend to be overconfident about what they know in ordinary situations such as correctly identifying facts about the physical world (Myers, 2002). Cults are made up of people convinced of the validity of their beliefs, for all that such beliefs are discordant with those of the remainder of society (Barušs, 1996). Conviction functions as a reasoning bias, so that it is only prudent to revisit this question of validity.

There is a fundamental sense in which conviction stems from implicit cognition. Alongside the contents of consciousness, at least within the dualistic realm, are *feelings of reality* and *feelings of knowing* regarding those contents. Such feelings are themselves seldom the focus of attention, nor could they ever perhaps be fully cognized as such, but rather provide part of the phenomenological context within which contents occur. Feelings of reality can be diminished, as they are in pathological states of derealization (American Psychiatric Association, 2000), and they can be enhanced, in cases of drug-induced states, near-death experiences, and transcendent states, as we have already noted.

One of my students, Lee-Anna Sangster, tried to empirically find some of the characteristics of feelings of reality in both university students and those who had had near-death experiences. The strategy was to direct participants' attention to their feelings of reality in both ordinary and altered states so as to see if they could identify the features of feelings of reality. On the basis of observations from 15 participants, she found that feelings of reality were more accessible, both in terms of clarity and intensity, as well as being more distinguishable, in altered states of consciousness than they were in the ordinary waking state. Furthermore, it appears that, in the ordinary waking state, participants used judgments about the objective nature of sensory objects to draw conclusions about what was real, whereas feelings of reality played a greater role in the altered states (Sangster, 2004).

The fact that characteristics of feelings of reality could be found that changed between ordinary and altered states, confirms that feelings of reality are not stable and suggests that they may be affected by factors other than the actual ontological status of whatever is being experienced. In other words, feelings of reality may well become more salient in transcendent states but meaningless as a criterion of validity. Or perhaps it is the other way around. The salience of feelings of reality in altered states testifies to their value for making judgments about reality in contrast to the apparently automatic evaluations about the objective nature of sensory objects.

It is not clear to what extent such tentative findings about feelings of reality are relevant to knowledge in transcendent states of consciousness. After all, we do not know in what way implicit cognition is carried over into such states. Furthermore, as we have already seen, according to Wolff, the feelings of reality associated with introception are not sensory impressions, given that they result from identification with that which is known, but have somewhat the quality of the logical inevitability of mathematical proofs. In this sense, they are closer in character to the feelings of reality in near-death experiences that are associated with transcendental awareness. And the nature of validity itself could be radically changed in transcendent states. Just as I realized, with the occurrence of my image of the flower stretching above the cube, as described in Part II, the meaningfulness of inner knowledge could outgrow the boundaries of concerns about validity. And it might not be until we have reached a particular point in our self-transformations that we are able to resolve questions of validity in some manner that seems appropriate at the time.

But perhaps also we can simply apply the criterion of validity that we established in Parts I and II whereby we conceptualized understanding as a synthetic process of dialectical reasoning. It seems to me that, to the extent that inner knowledge in transcendent states of consciousness further unfolds understanding, the same arguments concerning validity would apply. We know everything that we knew previously plus we know more. That was the impression that I had during a transient transcendent experience that occurred for me sometime around 1972.

In the mystical experience that I had, I spontaneously entered a transcendent state of consciousness in which I had the sense that I knew everything that I ordinarily knew, but saw that it all fit together in a harmonious pattern with everything else in the universe. Coincident with that noetic realization I had a deep sense of emotional well-being that was not characterized by excitement, but rather by profound groundedness. This state of being lasted for about 5 to 10 minutes and did not affect my outward behaviour, which consisted of interacting in a conversation about completely unrelated matters with a couple of friends.

At the time of my experience, it was clear to me that my understanding was being supplemented by insights that were normally unavailable to me. The impression was one of having become liberated from a constricted view of reality into a more expanded domain. In retrospect, after years of conventional scientific indoctrination, it is easy for me to question the validity of any insights that I had at the time. The same doubts do not pertain to the feelings that were part of this occurrence. Bliss is bliss if it is experienced as such, without concern for objective judgments about it. And perhaps enlightenment can be enlightenment if only for the person for whom it occurs. In the end, of course, we are back to the experimental method, this time to determine the nature of validity in transcendent states of consciousness.

In this part of the book we have been seeking transcendence. I started by describing the transcendent states of consciousness that occurred for Wren-Lewis and Wolff. According to Wolff, introception, knowledge through identity with that which is known, becomes activated in transcendent states. Theoretical thinking, when sufficiently abstract as in the case of mathematical thinking, can encompass some degree of introception, so that such transcriptive thinking can be used as a means of seeking transcendence. Mathematical yoga consists of doing mathematics in a meaningful way with abandonment of oneself to reality. Logical and mathematical invariants could play a role in self-transformation, as could the intention to imperience transcendent states of consciousness. And we need to experiment in order to develop our understanding and find out what works. In this way, science can be used as a spiritual practice.

Epilogue
Science as a Spiritual Practice

Existential questions can arise for scientists just as they can for anyone. When that occurs, I think that scientists should exploit the skills that they have already acquired rather than feeling that those are the wrong capabilities for spiritual aspiration. The question then becomes one of how scientific expertise can be reoriented toward spiritual aspiration.

We have already discussed some of the characteristics of scientific activity, such as that of sacrifice, that are consistent with spiritual effort. There are others that are self-evident. Many of these have been noted by Piero Ferrucci in describing science as a spiritual path. Curiosity can compel a scientist to ask troubling questions. Intense seeking of knowledge, the essence of authentic science, requires concentration, discipline, and intellectual honesty. A scientist has to eliminate entrenched beliefs in favour of seeing reality as it is. Although starting from that which is tangible, a scientist is often forced to leave behind a familiar world and to encounter unexpected aspects of nature. In some cases, this can lead to a radical change of her ideas and lifestyle so that she comes to appreciate the possibility of transcendence. These features of scientific engagement are all aspects of self-transformation (Ferrucci, 1990).

In addition to the incidental aspects of science that can contribute to spiritual development, we have discussed the use of abstract thinking as a way of approaching transcendent states of consciousness. According to Wolff, mathematical thinking, done in a meaningful manner and accompanied by a self-giving attitude, can be effective. But *perhaps understanding itself, oriented toward the root of reality, can reach a critical point of synthesis resulting in a radical transformation of consciousness and activation of latent ways of knowing.*

Of course, for scientists, such contentions are hypotheses to be empirically tested. Does open-minded meaningful mathematical reasoning aimed at transcendence work? Does the development of understanding lead to enlightenment? Which parameters are necessary for radical transformations of consciousness to occur? We can use our ingenuity to experiment. Thus, this book is not an answer but a challenge. How can science be used as a spiritual practice?

Endnotes

1. Borg, Andrée, Soderstrom, & Farde, 2003, in a study with 15 male participants, found that the apprehension of phenomena that cannot be objectively demonstrated covaries with low binding potential of the neurotransmitter serotonin to 5-HT$_{1A}$ receptor sites.

2. Suppose that a pair of photons is created and that one of them encounters a Polaroid filter set at angle A whereas the other encounters a filter set at angle B. Then suppose that we start again with another pair of photons and this time set the filters at angles A and C. And then do it again setting the filters at angles B and C. And then continue to send out pairs of polarized photons and to set filters at these three combinations of angles.

    If we were to do this experiment and assume that the photons carry the necessary information with them, then the number of photons that pass through at angle A and are blocked at angle B, plus the number of photons that pass through at angle B and are blocked at angle C, must be at least as many as the number that pass through at angle A and are blocked at angle C. In other words, the following, namely Bell's Theorem, must be true:

    $$n(A \sqcap \neg B) + n(B \sqcap \neg C) \geq n(A \sqcap \neg C)$$

    where "n" is the number of photons for the conditions in parentheses, "⊓" means "and", and "¬" means "not." This can be seen immediately to be true upon expansion of each of the three terms in the equation as follows:

    $$n(A \sqcap \neg B) = n(A \sqcap \neg B \sqcap C) + n(A \sqcap \neg B \sqcap \neg C)$$
    $$n(B \sqcap \neg C) = n(A \sqcap B \sqcap \neg C) + n(\neg A \sqcap B \sqcap \neg C)$$

    and

    $$n(A \sqcap \neg C) = n(A \sqcap B \sqcap \neg C) + n(A \sqcap \neg B \sqcap \neg C).$$

    In other words, substituting their expansions for each of the three terms in Bell's inequality gives:

$$\{n(A \cap \neg B \cap C) + n(A \cap \neg B \cap \neg C)\} + \{n(A \cap B \cap \neg C) + n(\neg A \cap B \cap \neg C)\}$$
$$\geq \{n(A \cap B \cap \neg C) + n(A \cap \neg B \cap \neg C)\}$$

which is clearly true. If there is a plan, whatever it may be, so that each photon of the pair knows what to do when it encounters polarizing filters set at specific angles, then, at a minimum, Bell's Theorem must be true.

This derivation is based loosely on Walker, 2000. The original paper is Bell, 1964/1983.

3   Malus' cosine-squared law states, in effect, that the number of photons that can pass through a Polaroid filter set at an angle to another one is proportional to the squared cosine of the angle between the filters. That means that the number stopped must be proportional to the squared sine.

With regard to the situation described in the previous endnote, we can suppose, as an example, that angle A is 0°, angle B is 30°, and angle C is 60°. If "E" represents the magnitude of the energy of the photons, then, using Malus' cosine-squared law:

$$n(A \cap \neg B) + n(B \cap \neg C) = E^2 \sin^2 30° + E^2 \sin^2 30° = E^2 \times \tfrac{1}{4} + E^2 \times \tfrac{1}{4} = E^2 \times \tfrac{1}{2}$$

and

$$n(A \cap \neg C) = E^2 \sin^2 60° = E^2 \times \tfrac{3}{4}$$

As a result, Bell's inequality becomes:

$$E^2 \times \tfrac{1}{2} \geq E^2 \times \tfrac{3}{4}$$

or:

$$\tfrac{1}{2} \geq \tfrac{3}{4}$$

which is not true, and Bell's inequality would be violated.
This discussion is based loosely on Walker, 2000.

4   Based loosely on Barušs, 1996; Polkinghorne, 2002; Feynman, Leighton, & Sands, 1965; Kafatos & Nadeau, 1990; Sudbery, 1986.

5   Qualities of Performance IQ summarized from Jackson, 1998.

6   Philosophical behaviourism, the idea that we are nothing but behaviour, has disappeared, but not methodological behaviourism, the idea that the proper object of observation in a psychological investigation is behaviour (Barušs, 1990a).

7   Notably Jahn & Dunne, 1986; Dunne, Jahn, & Nelson, 1985; Nelson, Dobyns, Dunne, & Jahn, 1991; also Jahn & Dunne, 1987; for a recent report see Jahn et al., 2000.

8   Based on Walker, 1970; 2000; Goswami, 1993; Barušs, 1986; see also Kafatos & Nadeau, 1990.

9   For a discussion of determinism and free will see Libet, Freeman, & Sutherland, 1999.

10 Cowan, O'Connor, & Samella, 2003; the calendrical savant was tested for five dates within the period 2,051,275 to 2,051,279.

11 What I assume now to have been the hamulus of the medial plate of pterygoid process; see e.g., Netter, 1986, p. 4.

12 Wren-Lewis, 1988, 1991, 1994; personal information based on material provided by Wren-Lewis in correspondence with the author, December 10, 2002.

13 Comment made to the author by Ron Leonard, who lived with Wolff during the last year of his life.

14 This is my transcription of the passage on the audiotape of Lecture #3 in Wolff's 1966 lecture series "Mathematics, Philosophy, and Yoga" (Merrell-Wolff, 1966). For the standard transcription see Merrell-Wolff, 1995a, p. 27.

15 A transcendental number is a real number that cannot be expressed as the root of a polynomial equation with integral coefficients (D. Nelson, 2003, p. 6; emphases removed in quotation).

Recommended Reading

The following are books that I think could be worthwhile as further reading about some of the subjects covered in this book. I have confined myself to a single selection for each topic. Readers can find other sources for the material that interests them by consulting the citations and endnotes which will direct them to the relevant entries in the list of references.

Science and Spirituality

Barušs, I. (1996). *Authentic knowing: The convergence of science and spiritual aspiration*. West Lafayette, IN: Purdue University Press.

Anomalous Phenomena

Radin, D. I. (1997). *The conscious universe: The scientific truth of psychic phenomena*. New York: HarperEdge.

Religious Experiences

Spilka, B., Hood, R. W., Jr., Hunsberger, B., & Gorsuch, R. (2003). *The psychology of religion: An empirical approach* (3rd ed.). New York: Guilford.

Self-Transformation

Ferrucci, P. (1982). *What we may be: Techniques for psychological and spiritual growth through psychosynthesis*. Los Angeles: Jeremy P. Tarcher.

Intuition

Myers, D. G. (2002). *Intuition: Its powers and perils*. New Haven, CT: Yale University Press.

Altered States of Consciousness

Barušs, I. (2003). *Alterations of consciousness: An empirical analysis for social scientists*. Washington, DC: American Psychological Association.

Dreams

Barrett, D. (2001). *The committee of sleep: How artists, scientists, and athletes use dreams for creative problem-solving – and how you can too*. New York: Crown.

Mathematical Yoga

Merrell-Wolff, F. (1995). *Mathematics, philosophy & yoga: A lecture series presented at the Los Olivos Conference Room in Phoenix, Arizona, in 1966*. Phoenix, AZ: Phoenix Philosophical Press.

References

Adler, S. L. (2003). Why decoherence has not solved the measurement problem: A response to P.W. Anderson (quant-ph/0112095). *Studies in History and Philosophy of Modern Physics, 34*, 135–142.

Allport, G.W., & Ross, J.M. (1967). Personal religious orientation and prejudice. *Journal of Personality and Social Psychology, 5* (4), 432-443.

American Psychiatric Association (2000). *Diagnostic and statistical manual of mental disorders* (4th ed. text revision). Washington, DC: Author.

Andresen, J. (2000). Meditation meets behavioural medicine: The story of experimental research on meditation. *Journal of Consciousness Studies, 7* (11–12), 17–73.

Appelle, S., Lynn, S.J., & Newman, L. (2000). Alien abduction experiences. In E. Cardeña, S.J. Lynn, & S. Krippner (Eds.), *Varieties of anomalous experience: Examining the scientific evidence* (pp. 253–282). Washington, DC: American Psychological Association.

Aspect, A., Dalibard, J., & Roger, G. (1982). Experimental test of Bell's inequalities using time-varying analyzers. *Physical Review Letters, 49*(25), 1804–1807.

Assagioli, R. (1965). *Psychosynthesis: A manual of principles and techniques*. New York: Penguin.

Assagioli, R. (1974). *The act of will*. Harmondsworth, England: Penguin. (Original work published 1973)

Assagioli, R. (1991). *Transpersonal development: The dimension beyond psychosynthesis*. London: Crucible. (Original work published 1988)

Bailey, A.A. (1950). *Letters on occult meditation*. New York: Lucis. (Original work published 1922)

Baker, D. (1975). *Meditation (the theory and practice)*. Essendon, England: Author.

Baker, D. (1977). *The spiritual diary*. Potters Bar, England: College of Spiritual Enlightenment and Esoteric Knowledge.

Barber, T.X. (1999). A comprehensive three-dimensional theory of hypnosis. In I. Kirsch, A. Capafons, E. Cardeña-Buelna, & S. Amigó (Eds.), *Clinical hypnosis and self-regulation: Cognitive-behavioral perspectives* (pp. 21–48). Washington, DC: American Psychological Association.

Barrett, D. (1993). The "committee of sleep": A study of dream incubation for problem solving. *Dreaming, 3*(2), 115-122.
Barrett, D. (2001). *The committee of sleep: How artists, scientists, and athletes use dreams for creative problem-solving – and how you can too.* New York: Crown.
Barrow, J.D. (1992). *Pi in the sky: Counting, thinking, and being.* Oxford: Clarendon.
Barušs, I. (1986). Quantum mechanics and human consciousness. *Physics in Canada, 42*(1), 3-5.
Barušs, I. (1987). Metanalysis of definitions of consciousness. *Imagination, Cognition and Personality, 6*(4), 321-329.
Barušs, I. (1989). Categorical modelling of Husserl's intentionality. *Husserl Studies, 6,* 25–41.
Barušs, I. (1990a). *The personal nature of notions of consciousness: A theoretical and empirical examination of the role of the personal in the understanding of consciousness.* Lanham, Maryland: University Press of America.
Barušs, I. (1990b). [Review of *Consciousness* by W.G. Lycan]. *Imagination, Cognition and Personality, 9*(2), 179–182.
Barušs, I. (1992). [Review of *A cognitive theory of consciousness* by B.J. Baars]. *Imagination, Cognition and Personality, 11*(3), 303–307.
Barušs, I. (1993). Can we consider matter as ultimate reality? Some fundamental problems with a materialist interpretation of reality. *Ultimate Reality and Meaning: Interdisciplinary Studies in the Philosophy of Understanding, 16*(3–4), 245–254.
Barušs, I. (1995). [Review of *The psychology of consciousness* by G.W. Farthing]. *Imagination, Cognition and Personality, 14*(4), 353–355.
Barušs, I. (1996). *Authentic knowing: The convergence of science and spiritual aspiration.* West Lafayette, IN: Purdue University Press.
Barušs, I. (1998). [Review of *Consciousness reconsidered* by O. Flanagan]. *Imagination, Cognition and Personality, 18*(2), 167–170.
Barušs, I. (2000). Psychopathology of altered states of consciousness. *Journal of Baltic Psychology, 1*(1), 12–26.
Barušs, I. (2003). *Alterations of consciousness: An empirical analysis for social scientists.* Washington, DC: American Psychological Association.
Barušs, I., & Moore, R. J. (1998). Beliefs about consciousness and reality of participants at 'Tucson II'. *Journal of Consciousness Studies, 5*(4), 483–496.
Batson, C.D., Schoenrade, P., & Ventis, W.L. (1993). *Religion and the individual: A social-psychological perspective.* New York: Oxford University Press. (Original work published 1982)
Bauer, H.H. (1992). *Scientific literacy and the myth of the scientific method.* Urbana, IL: University of Illinois Press.
Bell, J.L. (1977). *Boolean-valued models and independence proofs in set theory.* Oxford, England: Clarendon.

References

Bell, J.S. (1983). On the Einstein Podolsky Rosen paradox. In J.A. Wheeler & W.H. Zurek (Eds.), *Quantum theory and measurement* (pp. 403-408). Princeton, NJ: Princeton University Press. (Original work published 1964)

Bibby, R.W. (1987). *Fragmented gods: The poverty and potential of religion in Canada.* Toronto, Canada: Irwin.

Bierman, D.J., & Radin, D. (1999). Conscious and anomalous nonconscious emotional processes: A reversal of the arrow of time? In S.R. Hameroff, A.W. Kaszniak, & D.J. Chalmers (Eds.), *Toward a Science of Consciousness III: The third Tucson discussions and debates* (pp. 367-385). Cambridge, MA: MIT Press.

Bohm, D. (1983). *Wholeness and the implicate order.* London: Ark. (Original work published 1980)

Borg, J, Andrée, B., Soderstrom, H., & Farde, L. (2003). The serotonin system and spiritual experiences. *American Journal of Psychiatry, 160*(11), 1965-1969.

Bowker, J. (Ed.). (1997). *The Oxford dictionary of world religions.* Oxford, England: Oxford University Press.

Braude, S.E. (2003). *Immortal remains: The evidence for life after death.* Lanham, MD: Rowman & Littlefield.

Bray, F. (1980). *The philosopher's stone* [interview with Franklin Wolff at Lone Pine, California]. [Motion picture].

Brown, C. (1996). *Cosmic voyage: A scientific discovery of extraterrestrials visiting earth.* New York: Dutton.

Brown, D., Scheflin, A.W., & Hammond, D.C. (1998). *Memory, trauma treatment, and the law.* New York: W. W. Norton.

Brown, M.F. (1997). *The channeling zone: American spirituality in an anxious age.* Cambridge, MA: Harvard University Press.

Burch, W.E. (2003). *She who dreams: A journey into healing through dreamwork.* Novato, CA: New World Library.

Caird, D. (1987). Religiosity and personality: Are mystics introverted, neurotic, or psychotic? *British Journal of Social Psychology, 26,* 345-346.

Cann, D.R., & Harris, J.A. (2003). Assessment of the curious experiences survey as a measure of dissociation. *Personality and Individual Differences, 35,* 489-499.

Cardeña, E., Lynn, S.J., & Krippner, S. (Eds.). (2000). *Varieties of anomalous experience: Examining the scientific evidence.* Washington, DC: American Psychological Association.

Chamberlain, D.B. (1986). Reliability of birth memory: Observations from mother and child pairs in hypnosis. *Journal of the American Academy of Medical Hypnoanalysts, 1,* 89-98.

Child, I.L. (1985). Psychology and anomalous observations: The question of ESP in dreams. *American Psychologist, 40* (11), 1219-1230.

Cialdini, R.B. (1988). *Influence: Science and practice* (2nd ed.). New York: HarperCollins. (Original work published 1985)

Cowan, R., O'Connor, N., & Samella, K. (2003). The skills and methods of calendrical savants. *Intelligence, 31,* 51-65.

d'Aquili, E.G., & Newberg, A.B. (1999). *The mystical mind: Probing the biology of religious experience*. Minneapolis, MN: Fortress.

d'Aquili, E.G., & Newberg, A.B. (2000). The neuropsychology of aesthetic, spiritual, and mystical states. *Zygon, 35*(1), 39-51.

Dawson, L.L. (1987). On references to the transcendent in the scientific study of religion: A qualified idealist proposal. *Religion, 17*, 227-250.

De Koninck, J. (2000). Waking experiences and dreaming. In M.H. Kryger, T. Roth, & W.C. Dement (Eds.), *Principles and practice of sleep medicine* (3rd ed., pp. 502-509). Philadelphia: W.B. Saunders.

Decuypere, J. -M. (1999). Channelling — sick or scientific? *Journal of the Society for Psychical Research, 63* (856), 193-202.

Deleuze, G. (1994). *Difference and repetition* (P. Patton, Trans.). New York: Columbia University Press. (Original work published 1968) [epigraph quoted from page xxi]

Dement, W.C. (1978). *Some must watch while some must sleep*. New York: W.W. Norton. (Original work published 1972)

Dennett, D.C. (1978). *Brainstorms: Philosophical essays on mind and psychology*. Montgomery, VT: Bradford.

Dennett, D.C. (1988). Quining qualia. In A.J. Marcel & E. Bisiach (Eds.), *Consciousness in contemporary science* (pp. 42-77). Oxford, England: Clarendon.

Doblin, R. (1991). Pahnke's "Good Friday experiment": A long-term follow-up and methodological critique. *Journal of Transpersonal Psychology, 23* (1), 1-28.

Dodwell, P. (2000). *Brave new mind: A thoughtful inquiry into the nature and meaning of mental life*. New York: Oxford University Press.

Dorfman, J., Shames, V.A., & Kihlstrom, J.F. (1996). Intuition, incubation, and insight: Implicit cognition in problem solving. In G. Underwood (Ed.), *Implicit cognition* (pp. 257-296). Oxford: Oxford University Press.

Dreyfus, H.L., & Dreyfus, S.E. (with Athanasiou, T.). (1986). *Mind over machine: The power of human intuition and expertise in the era of the computer*. New York: The Free Press.

Dunne, B.J., Jahn, R.G., & Nelson, R.D. (1985). *Princeton Engineering Anomalies Research* (Technical Note PEAR 85003). Princeton, NJ: Engineering Anomalies Research Laboratory, School of Engineering/Applied Science, Princeton University.

Edward, J. (1998). *One last time: A psychic medium speaks to those we have loved and lost*. New York: Berkeley.

Edward, J. (2001). *Crossing over: The stories behind the stories*. San Diego: Jodere Group.

Edward, J. (with N. Stoynoff). (2003). *After life: Answers from the other side*. New York: Princess.

Emmons, R.A. (1999). *The psychology of ultimate concerns: Motivation and spirituality in personality*. New York: Guilford.

Emmons, R.A., & Paloutzian, R.F. (2003). The psychology of religion. *Annual Review of Psychology, 54*, 377-402.

Fenwick, P., & Fenwick, E. (1995). *The truth in the light: An investigation of over 300 near-death experiences.* New York: Berkeley.

Ferrucci, P. (1982). *What we may be: Techniques for psychological and spiritual growth through psychosynthesis.* Los Angeles: Jeremy P. Tarcher.

Ferrucci, P. (1990). *Inevitable grace: Breakthroughs in the lives of great men and women: Guides to your self-realization* (D. Kennard, Trans.). Los Angeles: Jeremy P. Tarcher.

Feynman, R.P., Leighton, R.B., & Sands, M. (1965). *The Feynman lectures on physics: Quantum mechanics.* Reading, MA: Addison-Wesley.

Fiedler, K. (2000). On mere considering: The subjective experience of truth. In H. Bless and J.P. Forgas (Eds.) *The message within: The role of subjective experience in social cognition and behavior* (pp. 13-36). Philadelphia, PA: Psychology Press.

Firman, J., & Gila, A. (2002). *Psychosynthesis: A psychology of the spirit.* Albany, NY: State University of New York Press.

Fodor, J. (2000). *The mind doesn't work that way: The scope and limits of computational psychology.* Cambridge, MA: MIT Press.

Frankel, F.H., & Covino, N.A. (1997). Hypnosis and hypnotherapy. In P.S. Appelbaum, L.A. Uyehara, & M.R. Elin (Eds.), *Trauma and memory: Clinical and legal controversies* (pp. 344-359). New York: Oxford University Press.

Freedman, T.B. (2002). *Soul echoes: The healing power of past-life therapy.* New York: Citadel.

Gardner, H. (1983). *Frames of mind: The theory of multiple intelligences.* New York: Basic.

Gardner, H. (2000). A case against spiritual intelligence. *International Journal for the Psychology of Religion, 10*(1), 27-34.

Goswami, A. (with Reed, R.E., & Goswami, M.). (1993). *The self-aware universe: How consciousness creates the material world.* New York: Jeremy P. Tarcher / Putnam.

Grabhorn, L. (2000). *Excuse me, your life is waiting: The astonishing power of feelings.* Charlottesville, VA: Hampton Roads Publishing Company.

Graff, D.E. (2000). *River dreams: The case of the missing general and other adventures in psychic research.* Boston, MA: Element.

Greyson, B. (2000). Near-death experiences. In E. Cardeña, S.J. Lynn, & S. Krippner (Eds.), *Varieties of anomalous experience: Examining the scientific evidence* (pp. 315-352). Washington, DC: American Psychological Association.

Grzegorczyk, A. (1964). A philosophically plausible formal interpretation of intuitionistic logic. *Indagationes Mathematicae, 26,* 596-601.

Hardy, J. (1987). *A psychology with a soul: Psychosynthesis in evolutionary context.* London, England: Arkana.

Harman, W., & Rheingold, H. (1984). *Higher creativity: Liberating the unconscious for breakthrough insights.* Los Angeles: Jeremy P. Tarcher.

Hastings, A. (1988). Exceptional abilities in channeling. *Noetic Sciences Review, 6,* 27–29.

Hastings, A. (1991). *With the tongues of men and angels: A study of channeling.* Fort Worth, TX: Holt, Rinehart and Winston.

Heaton, P., & Wallace, G.L. (2004). Annotation: The savant syndrome. *Journal of Child Psychology and Psychiatry, 45*(5), 899–911.

Heidegger, M. (1962). *Being and time* (J. Macquarrie & E. Robinson, Trans.). New York: Harper & Row. (Original work published 1926)

Heyting, A. (1976). Intuitionism: An introduction (3rd rev. ed.). Amsterdam: North-Holland. (Original work published 1956)

Hobson, J.A. (1988). *The dreaming brain.* New York: Basic.

Hobson, J.A. (1990). Dreams and the brain. In S. Krippner (Ed.), *Dreamtime and dreamwork: Decoding the language of the night* (pp. 215–223). Los Angeles: Jeremy P. Tarcher.

Hobson, J.A., & McCarley, R.W. (1977). The brain as a dream state generator: An activation-synthesis hypothesis of the dream process. *American Journal of Psychiatry, 134*(12), 1335–1348.

Hobson, J.A., Pace-Schott, E.F., & Stickgold, R. (2000). Consciousness: Its vicissitudes in waking and sleep. In M. S. Gazzaniga (Ed. in chief), *The new cognitive neurosciences* (2nd ed. pp. 1341–1354). Cambridge, MA: MIT Press.

Hogarth, R.M. (2001). *Educating intuition.* Chicago: University of Chicago Press.

Hood, R.W., Jr., Spilka, B., Hunsberger, B., & Gorsuch, R. (1996). *The psychology of religion: An empirical approach* (2nd ed.). New York: Guilford.

Horgan, J. (2003). *Rational mysticism: Dispatches from the border between science and spirituality.* Boston: Houghton Mifflin.

Huber-Dyson, V. (1991). *Gödel's theorems; a workbook on formalization.* Stuttgart, Germany: B.G. Teubner Verlagsgesellschaft.

Hughes, D.J. (1992). Differences between trance channeling and multiple personality disorder on structured interview. *The Journal of Transpersonal Psychology, 24*(2), 181–192.

Jackson, D.N. (1998), *Multidimensional Aptitude Battery-II: Manual.* Port Huron, MI: Sigma Assessment Systems.

Jackson, D.N. (1999). *Personality Research Form: Manual* (3rd ed). Port Huron, MI: Sigma Assessment Systems.

Jacobs, D.M. (Ed.) (2000). *UFOs and abductions: Challenging the borders of knowledge.* Lawrence, KS: University Press of Kansas.

Jahn, R.G. (2001). 20th and 21st century science: Reflections and projections. *Journal of Scientific Exploration, 15*(1), 21–31.

Jahn, R.G., & Dunne, B.J. (1986). On the quantum mechanics of consciousness, with application to anomalous phenomena. *Foundations of Physics, 16*(8), 721–772.

Jahn, R.G., & Dunne, B.J. (1987). *Margins of reality: The role of consciousness in the physical world.* San Diego, CA: Harcourt Brace Jovanovich.

Jahn, R., Dunne, B., Bradish, G., Dobyns, Y., Lettieri, A., Nelson, R., Mischo, J., Boller, E., Bösch, H., Vaitl, D., Houtkooper, J., & Walter, B. (2000). Mind/machine interaction consortium: PortREG replication experiments. *Journal of Scientific Exploration, 14* (4), 499–555.

James, W. (1904a). A world of pure experience. I. *Journal of Philosophy, Psychology and Scientific Methods, 1* (20), 533–543.

James, W. (1904b). A world of pure experience. II. *Journal of Philosophy, Psychology and Scientific Methods, 1* (21), 561–570.

James, W. (1958). *The varieties of religious experience.* New York: Mentor. (Original work published 1902)

Jeans, J. (1937). *The mysterious universe* (Rev. ed.). New York: Macmillan. (Original work published 1930)

Jewkes, S., & Barušs, I. (2000). Personality correlates of beliefs about consciousness and reality. *Advanced Development: A Journal on Adult Giftedness, 9,* 91–103.

Johnson, G. (1990). Telepathic dreams. *Consciousness Review, 1,* 37–46.

Jung, C.G. (1971). *Psychological types.* In W. McGuire (Executive Ed.), H. Read, M. Fordham, & G. Adler (Eds.), H.G. Baynes (Trans.) & R. F. C. Hull (Revisor of English trans.), *The collected works of C.G. Jung: Vol. 6.* Princeton, NJ: Princeton University Press. (Original work without appendix published 1921)

Kafatos, M., & Nadeau, R. (1990). *The conscious universe: Part and whole in modern physical theory.* New York: Springer-Verlag.

Kellehear, A. (1996). *Experiences near death: Beyond medicine and religion.* New York: Oxford University Press.

Klimo, J. (1987). *Channeling: Investigations on receiving information from paranormal sources.* Los Angeles: Jeremy P. Tarcher.

Klinger, E. (1978). Modes of normal conscious flow. In K.S. Pope & J.L. Singer (Eds.), *The stream of consciousness: Scientific investigations into the flow of human experience* (pp. 225–258). New York: Plenum.

Klinger, E. (1990). *Daydreaming: Using waking fantasy and imagery for self-knowledge and creativity.* Los Angeles: Jeremy P. Tarcher.

Klinger, E., & Kroll-Mensing, D. (1995). Idiothetic assessment: Experience sampling and motivational analysis. In J.N. Butcher (Ed.), *Clinical personality assessment: Practical approaches,* (pp. 267–277). New York: Oxford University Press.

Krippner, S., & George, L. (1986). Psi phenomena as related to altered states of consciousness. In B.B. Wolman & M. Ullman (Eds.), *Handbook of states of consciousness* (pp. 332–364). New York: Van Nostrand Reinhold.

Krystal, J.H., Bennett, A., Bremner, J.D., Southwick, S.M., & Charney, D.S. (1996). Recent developments in the neurobiology of dissociation: Implications for posttraumatic stress disorder. In L.K.

Michelson & W.J. Ray (Eds.), *Handbook of dissociation: Theoretical, empirical, and clinical perspectives* (pp. 163-190). New York: Plenum.

LaBerge, S., & Rheingold, H. (1990). *Exploring the world of lucid dreaming*. New York: Ballantine.

Laing, D.R., & Cooper, D.G. (1971). *Reason & violence: A decade of Sartre's philosophy 1950-1960*. London: Tavistock. (Original work published 1964)

Lakoff, G., & Núñez, R.E. (1997). The metaphorical structure of mathematics: Sketching out cognitive foundations for a mind-based mathematics. In L.D. English (Ed.) *Mathematical reasoning: Analogies, metaphors, and images* (pp. 21-89). Mahwah, NJ: Lawrence Erlbaum Associates.

Lakoff, G., & Núñez, R.E. (2000). *Where mathematics comes from: How the embodied mind brings mathematics into being*. New York: Basic.

Leonard, D. (1995). Forward. In F. Merrell-Wolff (Author), *Mathematics, philosophy & yoga: A lecture series presented at the Los Olivos Conference Room in Phoenix, Arizona, in 1966* (p. v). Phoenix, AZ: Phoenix Philosophical Press.

Leonard, R. (1999). *The transcendental philosophy of Franklin Merrell-Wolff*. Albany, NY: State University of New York Press.

Leskowitz, E.D. (2000). Channeling and hypnosis. In E.D. Leskowitz (Ed.) *Transpersonal hypnosis: Gateway to body, mind and spirit* (pp. 163-180). Boca Raton, FL: CRC Press.

Lewicki, P., Hill, T., & Czyzewska, M. (1992). Nonconscious acquisition of information. *American Psychologist, 47* (6), 796-801.

Libet, B., Freeman, A., & Sutherland, K. (1999). Editors' introduction: The volitional brain: Towards a neuroscience of free will. *Journal of Consciousness Studies, 6* (8-9), ix-xxii.

Lukey, N. & Barušs, I. (2005). Intelligence correlates of transcendent beliefs: A preliminary study. *Imagination, Cognition and Personality, 24* (3), 259-270.

Lukoff, D. (1985). The diagnosis of mystical experiences with psychotic features. *Journal of Transpersonal Psychology, 17* (2), 155-181.

Lukoff, D., & Everest, H.C. (1985). The myths in mental illness. *Journal of Transpersonal Psychology, 17* (2), 123-153.

Lunn, J.H. (1948). Chick sexing. *American Scientist, 36* (2), 280-287.

Mack, J.E. (1994). Why the abduction phenomenon cannot be explained psychiatrically. In A. Pritchard, D.E. Pritchard, J.E. Mack, P. Kasey, & C. Yapp (Eds.), *Alien discussions: Proceedings of the abduction study conference held at MIT, Cambridge, MA* (pp. 372-374). Cambridge, MA: North Cambridge.

Mack, J.E. (1999). *Passport to the cosmos: Human transformation and alien encounters*. New York: Three Rivers.

MacMartin, C., & Yarmey, A.D. (1999). Rhetoric and the recovered memory debate. *Canadian Psychology/Psychologie canadienne, 40* (4), 343-358.

Malmgren, J. (1994, November 27). Tune in, turn on, get well? *St. Petersburg Times*, p. 1F.

Mandelbaum, W.A. (2000). *The psychic battlefield: A history of the military-occult complex*. New York: St. Martin's.

Mandell, A.J. (1980). Toward a psychobiology of transcendence: God in the brain. In J.M. Davidson & R.J. Davidson (Eds.), *The psychobiology of consciousness* (pp. 379–464). New York: Plenum.

Mandler, G. (1985). *Cognitive psychology: An essay in cognitive science*. Hillsdale, NJ: Lawrence Erlbaum Associates.

Mavromatis, A. (1987). *Hypnagogia: The unique state of consciousness between wakefulness and sleep*. London: Routledge & Kegan Paul.

May, E.C. (1996). The American Institutes for Research Review of the Department of Defense's STAR GATE program: A commentary. *Journal of Scientific Exploration, 10* (1), 89–107.

McFarlane, A.C., & van der Kolk, B.A. (1996). Conclusions and future directions. In B.A. van der Kolk, A.C. McFarlane, & L. Weisaeth (Eds.), *Traumatic stress: The effects of overwhelming experience on mind, body, and society* (pp. 559–575). New York: Guilford.

McLeod, C.C., Corbisier, B., & Mack, J.E. (1996). A more parsimonious explanation for UFO abduction. *Psychological Inquiry, 7* (2), 156–168.

Merrell-Wolff, F. (Speaker). (1966). *Mathematics, philosophy and yoga*. (Cassette recordings of six lectures, 1966, Phoenix, Arizona). Phoenix, AZ: Phoenix Philosophical Press.

Merrell-Wolff, F. (Speaker). (1970). *The induction* (Cassette recording January 24, 1970, Phoenix, Arizona). Phoenix, AZ: Phoenix Philosophical Press.

Merrell-Wolff, F. (Speaker). (1975). *The high indifference*. (Cassette recording April 30, 1975, Lone Pine, California: Phoenix, AZ: Phoenix Philosophical Press.

Merrell-Wolff, F. (1994). *Franklin Merrell-Wolff's experience and philosophy: A personal record of transformation and a discussion of transcendental consciousness*. Albany, NY: State University of New York Press.

Merrell-Wolff, F. (1995a). *Mathematics, philosophy & yoga: A lecture series presented at the Los Olivos Conference Room in Phoenix, Arizona, in 1966*. Phoenix, AZ: Phoenix Philosophical Press.

Merrell-Wolff, F. (1995b). *Transformations in consciousness: The metaphysics and epistemology*. Albany, NY: State University of New York Press.

Mills, A., & Lynn, S.J. (2000). Past-life experiences. In E. Cardeña, S.J. Lynn, & S. Krippner (Eds.), *Varieties of anomalous experience: Examining the scientific evidence* (pp. 283–313). Washington, DC: American Psychological Association.

Moody, R.A., Jr. (1988). *The light beyond*. New York: Bantam.

Morehouse, D. (1996). *Psychic warrior: Inside the CIA's Stargate program: The true story of a soldier's espionage and awakening*. New York: St. Martin's.

Musès, C. (1970). Altering states of consciousness by mathematics, with applications to education. *Journal for the Study of Consciousness, 3* (1), 43-50.

Myers, D.G. (2002). *Intuition: Its powers and perils.* New Haven, CT: Yale University Press.

Nadel, L., & Jacobs, W.J. (1998). Traumatic memory is special. *Current Directions in Psychological Science, 7* (5), 154-157.

Nelson, D. (Ed.). (2003). *The Penguin dictionary of mathematics* (3rd ed.). London: Penguin.

Nelson, R. (2000). Experiment at the annual meeting: A beautifully scientific sunset. *The Explorer: Newsletter of the Society for Scientific Exploration, 16* (3), p. 5.

Nelson, R.D. (2002). Coherent consciousness and reduced randomness: Correlations on September 11, 2001. *Journal of Scientific Exploration, 16* (4), 549-570.

Nelson, R. (2003). Coherent consciousness and reduced randomness. *Journal of Scientific Exploration, 17* (2), 339-341.

Nelson, R.D., Dobyns, Y.H., Dunne, B.J., & Jahn, R.G. (1991). *Analysis of variance of REG experiments: Operator intention, secondary parameters database structure* (Technical Note PEAR 91004). Princeton, NJ: Princeton Engineering Anomalies Research, Princeton University.

Nelson, R.D., Dunne, B.J., Dobyns, Y.H., & Jahn, R.G. (1996). Precognitive remote perception: Replication of remote viewing. *Journal of Scientific Exploration, 10* (1), 109-110.

Nelson, R.D., Jahn, R.G., Dunne, B.J., Dobyns, Y.H., & Bradish, G.J. (1998). FieldREG II: Consciousness field effects: Replications and explorations. *Journal of Scientific Exploration, 12* (3), 425-454.

Netter, F.H. (1986). *The CIBA collection of medical illustrations. Volume 1: Nervous system. Part I: Anatomy and physiology.* West Caldwell, NJ: CIBA.

Northrop, F.S.C. (1966). *The meeting of East and West: An inquiry concerning world understanding.* New York: Collier. (Original work published 1946)

Pahnke, W.N. (1963). *Drugs and mysticism: An analysis of the relationship between psychedelic drugs and the mystical consciousness.* Unpublished doctoral dissertation, Harvard University.

Pargament, K.I. (1999). The psychology of religion *and* spirituality? Yes and no. *International Journal for the Psychology of Religion, 9* (1), 3-16.

Pekala, R.J., & Kumar, V.K. (2000). Operationalizing "trance" I: Rationale and research using a psychophenomenological approach. *American Journal of Clinical Hypnosis, 43* (2), 107-135.

Persinger, M.A. (1987). *Neuropsychological bases of God beliefs.* New York: Praeger.

Petitmengin-Peugeot, C. (1999). The intuitive experience. *Journal of Consciousness Studies, 6* (2-3), 43-77.

Plato. (1968). *The republic* (B. Jowett, Trans.). New York: Airmont.

Polkinghorne, J. (2002). *Quantum theory: A very short introduction*. Oxford, United Kingdom: Oxford University Press.

Puthoff, H.E. (1996). CIA-initiated remote viewing program at Stanford Research Institute. *Journal of Scientific Exploration, 10* (1), 63–76.

Radin, D.I. (1997a). *The conscious universe: The scientific truth of psychic phenomena*. New York: HarperEdge.

Radin, D.I. (1997b). Unconscious perception of future emotions: An experiment in presentiment. *Journal of Scientific Exploration, 11* (2), 163–180.

Radin, D. (2002). Exploring relationships between random physical events and mass human attention: Asking for whom the bell tolls. *Journal of Scientific Exploration, 16* (4), 533–547.

Radin, D.I. (2004). Electrodermal presentiments of future emotions. *Journal of Scientific Exploration, 18* (2), 253–273.

Rasiowa, H. (1974). Post algebras as a semantic foundation of m-valued logics. In A. Daigneault (Ed.), *Studies in Mathematics: Vol. 9. Studies in algebraic logic* (pp. 92–142). Washington: Mathematical Association of America.

Reed, H. (1976). Dream incubation: A reconstruction of a ritual in contemporary form. *Journal of Humanistic Psychology, 16* (4), 53–70.

Ring, K. (1987). Near-death experiences: Intimations of immortality? In J. S. Spong (Ed.), *Consciousness and survival: An interdisciplinary inquiry into the possibility of life beyond biological death* (pp. 165–176). Sausalito, CA: Institute of Noetic Sciences.

Ring, K., & Cooper, S. (1997). Near-death and out-of-body experiences in the blind: A study of apparent eyeless vision. *Journal of Near-Death Studies, 16* (2), 101–147.

Ring, K., & Valarino, E.E. (1998). *Lessons from the light: What we can learn from the near-death experience*. Portsmouth, NH: Moment Point.

Roberts, B. (1991). *The path to no-self: Life at the center*. Albany, NY: State University of New York Press.

Roberts, B. (1993). *The experience of no-self: A contemplative journey* (Rev. ed.). Albany, NY: State University of New York Press.

Roe, C.A. (1999). Critical thinking and belief in the paranormal: A re-evaluation. *British Journal of Psychology, 90*, 85–98.

Ross, C.A., Norton, G.R., & Wozney, K. (1989). Multiple personality disorder: An analysis of 236 cases. *Canadian Journal of Psychiatry, 34* (5), 413–418.

Sangster, L.T. (2004). *Characterizing feelings of reality*. Bachelor's thesis, University of Western Ontario, London, Ontario, Canada.

Sartre, J.-P. (1960). *Critique de la raison dialectique (précédé de questions de méthode)*. Gallimard.

Scargle, J.D. (2002). Was there evidence of global consciousness on September 11, 2001? *Journal of Scientific Exploration, 16* (4), 571–577.

Schacter, D.L. (1995). Memory distortion: History and current status. In D.L. Schacter (Ed.), *Memory distortion: How minds, brains, and societies reconstruct the past* (pp. 1–43). Cambridge, MA: Harvard University Press.

Schatzman, M. (1983a). Solve your problems in your sleep. *New Scientist, 98* (1361), 692–693.

Schatzman, M. (1983b). Sleeping on problems really can solve them. *New Scientist, 99* (1370), 416–417.

Schnabel, J. (1997). *Remote viewers: The secret history of America's psychic spies.* New York: Dell.

Schwartz, G.E. (with Simon, W.L.). (2002). *The afterlife experiments: Breakthrough scientific evidence of life after death.* New York: Pocket.

Schwartz, J.M., Stapp, H.P., & Beauregard, M. (2005). Quantum physics in neuroscience and psychology: A neuropsychological model of mind-brain interaction. *Philosophical Transactions of the Royal Society B, 360* (1458), 1309–1327.

Shannon, C.E., & Weaver, W. (1964). *The mathematical theory of communication.* Urbana, IL: University of Illinois Press.

Shanon, B. (2002). *The antipodes of the mind: Charting the phenomenology of the ayahuasca experience.* Oxford, England: Oxford University Press.

Shorto, R. (1999). *Saints and madmen: Psychiatry opens its doors to religion.* New York: Henry Holt and Company.

Smith, R.C. (1990). Traumatic dreams as an early warning of health problems. In S. Krippner (Ed.), *Dreamtime and dreamwork*, (pp. 224–232). Los Angeles: Jeremy P. Tarcher.

Spanos, N.P., & Moretti, P. (1988). Correlates of mystical and diabolical experiences in a sample of female university students. *Journal for the Scientific Study of Religion, 27* (1), 105–116.

Spilka, B., Hood, R.W., Jr., Hunsberger, B., & Gorsuch, R. (2003). *The psychology of religion: An empirical approach* (3rd ed.). New York: Guilford.

Stevens, A. (1995). *Private myths: Dreams and dreaming.* Cambridge, MA: Harvard University Press.

Stevenson, I. (1997a). *Reincarnation and biology: A contribution to the etiology of birthmarks and birth defects.* Westport, CT: Praeger.

Stevenson, I. (1997b). *Where reincarnation and biology intersect.* Westport, CT: Praeger.

Strassman, R.J. (2001). *DMT: The spirit molecule.* Rochester, VT: Park Street.

Strauch, I., & Meier, B. (1996). *In search of dreams: Results of experimental dream research.* Albany, NY: State University of New York Press. (Original work published in 1992)

Sudbery, A. (1986). *Quantum mechanics and the particles of nature: An outline for mathematicians.* Cambridge, England: Cambridge University Press.

Targ, R. (1996). Remote viewing at Stanford Research Institute in the 1970s: A memoir. *Journal of Scientific Exploration, 10* (1), 77–88.

Targ, R., & Katra, J. (1998). *Miracles of mind: Exploring nonlocal consciousness and spiritual healing.* Novato, CA: New World Library.

Tart, C.T. (1972). States of consciousness and state-specific sciences. *Science, 176* (4038), 1203–1210.

Tart, C.T. (2000). Investigating altered states on their own terms: State-specific sciences. In M. Velmans (Ed.), *Investigating phenomenal consciousness: New methodologies and maps* (pp. 255–278). Amsterdam, The Netherlands: John Benjamins.

Taylor, E. (1999). *Shadow culture: Psychology and spirituality in America.* Washington, DC: Counterpoint.

Thalbourne, M.A. (1998). Transliminality: Further correlates and a short measure. *Journal of the American Society for Psychical Research, 92*, 402–419.

Tobacyk, J., & Milford, G. (1983). Belief in paranormal phenomena: Assessment instrument development and implications for personality functioning. *Journal of Personality and Social Psychology, 44* (5), 1029–1037.

Ullman, M. (1999). Dreaming consciousness: More than a bit player in the search for answers to the mind/body problem. *Journal of Scientific Exploration, 13* (1), 91–112.

Ullman, M., & Krippner, S. (with Vaughan, A.). (1973). *Dream telepathy.* New York: Macmillan.

Underwood, G., & Bright, J.E.H. (1996). Cognition with and without awareness. In G. Underwood (Ed.), *Implicit cognition* (pp. 1–40). Oxford: Oxford University Press.

van der Kolk, B.A. (1996). Trauma and memory. In B.A. van der Kolk, A.C. McFarlane, & L. Weisaeth (Eds.), *Traumatic stress: The effects of overwhelming experience on mind, body, and society* (pp. 279–302). New York: Guilford.

Vaughan, F.E. (1979). Transpersonal psychotherapy: Context, content and process. *Journal of Transpersonal Psychology, 11*(2), 101–110.

Waldron, J.L. (1998). The life impact of transcendent experiences with a pronounced quality of *noesis*. *Journal of Transpersonal Psychology, 30* (2), 103–134.

Walker, E.H. (1970). The nature of consciousness. *Mathematical Biosciences, 7*, 131–178.

Walker, E.H. (2000). *The physics of consciousness: Quantum minds and the meaning of life.* Cambridge, MA: Perseus.

Wallas, G. (1926). *The art of thought.* New York: Harcourt, Brace.

Weyl, H. (1989). *The open world: Three lectures on the metaphysical implications of science.* Woodbridge, CT: Ox Box. (Original work published 1932)

Wick, D. (1995). *The infamous boundary: Seven decades of controversy in quantum physics.* Boston: Birkhäuser.

Wigner, E.P. (1983a). Remarks on the mind-body question. In J.A. Wheeler & W.H. Zurek (Eds.), *Quantum theory and measurement* (pp. 168–181). Princeton, NJ: Princeton University Press. (Original work published 1961)

Wigner, E.P. (1983b). Interpretation of quantum mechanics. In J.A. Wheeler & W.H. Zurek (Eds.), *Quantum theory and measurement* (pp. 260–314). Princeton, NJ: Princeton University Press. (Original work given as lectures in 1976)

Woods, K., & Baruss, I. (2004). Experimental test of possible psychological benefits of past-life regression. *Journal of Scientific Exploration, 18* (4), 597–608.

Wren-Lewis, J. (1988). The darkness of God: A personal report on consciousness transformation through an encounter with death. *Journal of Humanistic Psychology, 28* (2), 105–122.

Wren-Lewis, J. (1991). A reluctant mystic: God-consciousness not guru worship. *Self and Society, 19* (2), 4–11.

Wren-Lewis, J. (1994). Aftereffects of near-death experiences: A survival mechanism hypothesis. *Journal of Transpersonal Psychology, 26* (2), 107–115.

Wuthnow, R. (1998). *After heaven: Spirituality in America since the 1950s.* Berkeley, CA: University of California Press.

Yewchuk, C. (1999). Savant syndrome: Intuitive excellence amidst general deficit. *Developmental Disabilities Bulletin, 27* (1), 58–76.

Zohar, D., & Marshall, I. (2000). *SQ – Spiritual intelligence: The ultimate intelligence.* London: Bloomsbury.

Index

academia, 6
  academic community, 15
  research and literature, 16, 21, 32, 70
academic activity as form of meditation, 117-118
academics, 14-15
alien abduction experiences, 23, 27, 38-40, 120
alterations of consciousness. *See* consciousness: alterations of
anomalous phenomena (anomalies), 5, 14-15, 20, 27, 30, 37, 40-41, 69
  anomalous experience, 31, 86
  anomalous information transfer, 57, 60 (*see also* remote viewing)
  controversies concerning, 29
  during altered states, 69
  evidence for, 20, 27, 45, 57, 60, 65
  See also under specific types of anomalous phenomena
Assagioli, Roberto, 43, 111
authenticity, 33, 86-87
  See also inauthenticity; volition
autism, 55
ayahuasca, 21-22, 114

backward causation, 11
  See also quantum mechanics
Bailey, Alice, 117
Baker, Douglas, 73
Barrett, Deirdre, 79, 82
Barrow, John, 58-59
behaviourism, 28, 128 n.6
beliefs, 22, 57
  about consciousness and reality, 16-20, 32
  materialist, 16, 19, 28, 38, 45 (*see also* materialism)
  persistence of, 35, 56, 120, 125
  religious, 17
  transcendent, 16-21, 23, 32, 35, 36, 38, 41, 43, 45, 56
  See also material-transcendent dimension; world view
Bell's Theorem (inequality), 8-9, 42, 45, 59
  proof of, 127-128 n.2, 128 n.3
  See also quantum mechanics
Bessent, Malcolm, 86
beta-carbolines, 21
Bohm, David, 43
brain, 1, 37, 129 n.11
  hippocampus, 37
  imaging techniques, 1
  mechanisms corresponding to mystical experience, 37-38
  serotonin system, activity of, 1, 37, 127 n.1
Buddhism, 114
Burch, Wanda, 82

channelling, 49, 50, 58-61, 69, 88
  channelled material, 58-60
    source of, 59
  and dissociation, 58, 70
  veridical nature of, 50, 59, 60
    evidence for, 60
  See also dissociation; inner knowledge
CIA (Central Intelligence Agency), 64
cognition. *See* thinking
cognitive schemata, 7, 12, 54, 107, 113
cognitive science, 53, 54, 107-108
Cohen, Paul, 113
consciousness, 12, 16, 18, 20, 104
  after death, 17, 28
  alterations of, 27, 47-49, 50, 68-84 (*see also under specific alterations*)
    drug-induced, 21-23, 91-92, 121

consciousness (cont.)
  alterations of
    empirical testing in, 118, 121
    as improving access to inner knowledge, 69
  beliefs about, 19, 32
    personality correlates of, 19
    role of, 16
  creating reality, 42
  deep, 43-44, 103
  definitions of, 25, 103
  as emergent property, 17, 22
  as hidden variable, 42
  as information, 103, 104
  and meaning, 17, 103, 109, 116
  nonconscious, 54, 57, 59, 61, 66, 68 (*see also* implicit cognition)
    forms of, 22-23
      ontological status of, 23
  ordinary (dualistic), 44, 50, 92, 94, 101, 105, 108-110, 115-116, 121
    as illusion, 120
    intentional structure of, 95-96, 101
    persistence of, 110
    phenomenology of, 96, 100, 121 (*see also* feelings)
    relation to transcendent consciousness, 92, 94-98, 101, 109-111
  preconscious, 44
  primacy of, 17, 103
  and quantum mechanics, 41-42
  and reality, 19, 116, 118
  as sense of existence, 17, 100-101, 103
  and spirituality, 17
  as stream, 17, 42, 43-44, 61, 100 (*see* experiential stream)
  study of, 16, 29, 121
  subconscious, 44
  as subjective awareness, 43-44
  subjective features of, 17, 28, 103
  superconscious, 44, 49, 88
  transcendent, 36, 37, 43-45, 91-92, 96, (*see also* transcendent states of consciousness)
    as non-objective, 101
  as ultimate reality, 18, 116
  unconscious, 54
creativity, 87
cults, 120

daydreaming, 62, 69
Dennett, Daniel, 29
derealization, 121
determinism, 7, 12, 14, 53, 129 n.9
  and free will, 129 n.9
  interaction of particles, 7
dialectical reasoning, 24-25, 30-31, 45, 122 *See also* understanding
dissociation, 58, 60, 66, 69-71, 87
  degrees of, 69-70
  dissociative identity disorder, 40, 70
doctrine, 1, 15, 28, 33, 45
Dodwell, Peter, 108
dreams, 72-86, 88
  characteristics of meaningful, 84
  directing of dreamer's attention in, 75-78
  dream architect, 85, 86
  empirical study of, 75-80, 82, 84-86
  about health, 82-83
  imagery, 75-77, 80-84
  incubation of, 73-75, 77-81
  knowledge, as source of, 72-73, 75, 77, 80-82, 84-86
  lucid, 73-75
  meaningfulness of, 84-85, 88
  meaning of, 72-73, 79-86
  presentient, 86
    evidence for, 86
  recall of, 74, 79
  symbols in (*see* symbols)
  telepathy, 75-77
  therapeutic value of, 79
dualism, 17

Edward, John, 60
Emmons, Robert, 36
empirical studies, 27, 66, 118, 121-122, 123, 126
enlightenment, 2, 34, 38, 49, 93-98, 109, 112, 117-118, 119-123, 126
  awakening, 94, 96, 98, 101, 112, 117, 119-120
  reasons for seeking, 119-120
  See also mystical experiences; transcendent states of consciousness
epistemological relativism, 24
Euler's formula, 107-108
evidence, as impetus for change, 20, 27

Index

existential concerns, 7, 17, 32, 45, 89
  crisis, 34
  meaning of life, 22, 34
  questions 1, 32-33, 87, 125
    answered in transcendent states, 93
  and science, 34
experiences
  diabolical, 38
  vs. imperience, 96
  personal, 3, 27-30, 45
    differences in, 29
    learning from as scientific, 29
  religious, 37, 104
  that science cannot explain, 19-21
  scientific, 30, 45
  sensory, 18, 101, 116
  unusual, 21, 31
  See also under specific types of experiences
experiential stream, 17, 42, 43-44, 61-62, 65, 67, 100-101
  See also consciousness
extrasensory perception, 17, 21, 56, 57, 64, 66
  in savants, 57
  understanding, as one component of, 66
  See also remote viewing

Faraday, Ann, 91
feelings, 110, 123
  of knowing, 56, 121
  of reality, 121-122
    empirical testing of, 121-122
Ferrucci, Piero, 125

Gödel, Kurt, 113
Good Friday experiment, 22
Grabhorn, Lynn, 50-51, 62
gratitude, 35
guidance, 2, 49, 86-87
  See also inner knowledge
guided imagery, 62, 70
  See also inner knowledge; intuition

Heidegger, Martin, 33
human-machine interaction, 5-6, 30
humanity, 119
hypnagogic imagery, 67-69, 81
  autosymbolic nature of, 67, 81

hypnotism, 70-71
  empirical studies of, 71-2
  and false memories, 71
  past life regression, 72
  susceptibility to, 70-71

imagination, 62, 71, 88
imperience, 96
implicit cognition, 53, 54-57, 66, 108, 121
  cognitive mechanisms of, 57
  empirical studies of, 54-57
  extraordinary components of, 58, 66, 88
  as intuition, 57
  within transcendent states, 122
  as part of understanding, 66
  validity of, 56
  See also consciousness: nonconscious
inauthenticity, 33, 63-64, 87
  See also authenticity
information, 9-10, 28, 29-30, 31, 56, 58, 103-104
  flow of, 100
  nonsensory acquisition of, 49, 57-60, 65, 86
  role of, 96-97
  understanding, as component of, 104
  See also knowledge
inner growth, 18
inner knowledge, 2, 32, 45, 49, 52-54, 61, 63, 87-88, 98
  access to, 49-50, 70, 77, 86
  through dreams, 77-86
  empirical research about, 50, 55, 65, 67
  extraordinary source of, 58
  through hypnotic regression, 71
  as implicit cognition, 54, 57, 88
  primacy of, 49
  and rare events, 52
  as seeing, 99-100
  and spontaneous thinking, 61-64
  symbolic forms of, 68
  See also extrasensory perception; guidance; presentience
intelligence, 19, 35
  correlation with transcendent beliefs, 19-20, 35, 45
  multiple forms of, 35
  spiritual, 36-37, 45

introception, 98-100, 102-104, 109, 120, 122-123
  transcriptive thinking, 102, 104, 107, 111, 123
introspection, 29
intuition, 49-51, 61, 66
  as coincidence, 51-52, 88
  empirical studies of, 61
  explanations of, 50
  hunches, 51-52, 54, 57-58, 86, 88
  as implicit cognition, 57 (*see also* implicit cognition)
  improving functioning of, 54, 61-62, 65
  through incubation, 61, 65-66
  intuitive impressions, 51-52, 54, 55, 58
    dangers, 87
    unreliability of, 12, 53
  as originating from superconscious, 44
  phenomenology of, 50, 61-62, 65-66, 69
  symbolic forms of, 67-68 (*see also* symbols)
  validity of, 50-52, 62-64, 86-87, 123
  See also knowledge; understanding

James, William, 22
Jeans, Sir James, 12
Jewkes, Sonya, 19
Jungian analysis, 73

Klinger, Eric, 61
knowledge
  authentic, 27
  conviction about, 120-121
    as reasoning bias, 120
  criteria for validity of, 31, 62-63, 121-123
  cumulative structure of, 24-25
  through extraordinary means, 49, 63, 65-66, 98 (*see also under specific types*)
  through identity, 98-99, 117, 123 (*see also* introception)
  implicit, 54, 58
  as information, 31
  judgmental heuristics, 54
  memory, 54
  personal quality of, 25
  personal ways of knowing, 27-28

knowledge (cont.)
  reasoning, 49, 53
    biases, 120
    implicit vs. explicit, 54
    strategies, 54
  scientific, 16, 27, 30
  through sensory perception, 18, 49, 54, 98, 101, 103
    veracity of, 18
    sources of, 54
  transcendent 99, 101, 122-123
  validity of, 53, 62-3, 119, 120, 122, 123
  value of, 120
  See also understanding; inner knowledge; intuition; noesis

LaBerge, Stephen, 73-74
Lakoff, George, 107-108
Leonard, Doroethy, 97
Leonard, Ron, 129 n.13
Lewicki, Paul, 54-55
logic, 113-115
  alternate, 118
  law of excluded middle, 113-114
  predicate calculus, 113-114
    classical, 113
    intuitionistic, 114, 115
  See also mathematics; thinking
Lukey, Nicole, 19
Lukoff, David, 39

Mack, John, 23
Mandell, Arnold, 37
material-transcendent dimension, 16-17, 45, 108
  dualist position, 17
  materialist position, 17, 108
  transcendent position, 17, 19, 108, 116
materialism, 6, 7, 12, 14, 16, 18, 20, 28, 33-34
  challenges (alternatives) to, 9, 24, 40, 45, 93
  discouragement from challenging, 15
  inadequacy of, 7, 20, 25, 32, 41, 44
  materialist beliefs, 19, 33, 38, 45
    shift to transcendent beliefs, 20-21, 27
  persistence of, 2, 14, 16, 28, 45
  vs. transcendence, 16
  See also scientism

Index

materialists, 17, 22, 24, 115
mathematical yoga, 1, 94, 105, 109, 111-112, 117, 123
  for mathematicians, 111
  for those with a scientific perspective, 111
mathematics
  arithmetic, 113
  axiomatization, 113-114
  e, 107, 115
  as expression of transcendent reality, 105, 115
  invariants, 112-113, 115, 123
    ontological status of, 115-116
  materialist view of, 115
  mathematical thinking, 102, 107-109, 111, 116, 123, 125-126
    lack of empirical studies of, 108
    structure of, 108-109
  and meaning, 108, 109, 115, 116, 125
  nature of, 105-106, 115
  as path to transcendent reality, 94, 105-106, 109, 112
  perceptions of, 105-106
  π (pi), 100, 107, 112, 115-116
  and reality, 115-116, 122
  set theory, 113
  as state-specific science, 118
  transcendental numbers, 116, 129 n.15
  as truth, 105
  See also logic; mathematical yoga
matter, 6, 7, 12, 14, 44
  nature of, 7, 20
  as ontologically primitive, 6
  See also quantum mechanics
meaning
  and anomalous phenomena, 13, 14
  as aspect of beliefs, 17, 32
  depth of, 95, 102-104, 109
  existential, 22, 32, 34
  of mathematical constructions, 108, 109
  meaningfulness, 25, 30, 32, 63, 104, 116, 122, 123
    of dreams, 72, 79, 84-86
  semantic, 22, 95, 102-103
meditation, 35, 37, 47, 51, 65-66, 68-69, 88, 112, 116-117
mediumship. *See* channelling

Merrell-Wolff, Franklin. *See* Wolff, Franklin
mind, 40, 68, 94, 107-108
  computational approach to, 104
  intentionality, 95
  See also cognitive science; thinking
Moody, Raymond, 26
Moore, Robert, 16-17, 21, 29, 31-32, 43
Musès, Charles, 106
mystical experiences, 36, 37-38, 39, 44, 49, 92, 97, 119, 122
  See also transcendent states of consciousness

N,N-*dimethyltriptamine* (DMT), 21, 22
near-death experience (NDE), 21, 26, 36, 38, 91, 99, 121-122
Nelson, Roger, 5, 6, 13, 30, 57
  global consciousness project, 57
new age, 50
noesis, 30, 31, 88, 120
  noetic quality of transcendent experiences, 36, 49, 93, 120, 122
  See also knowledge; understanding
non-locality, 10
  See also quantum mechanics
Northrop, F.S.C., 106
Núñez, Rafael, 107-108

observation, 12, 28, 41
out-of-body experiences, 17, 99

paradigms, 2, 24
particles, 7, 10-11, 41-42, 45
philosophy, 93, 94, 109, 117
physicalism, 6-7, 12, 20
physical world, 18, 48
  vs. unmanifest world, 47-48
  See also reality
physics, 7, 18, 42
  See also quantum mechanics
Plato, 26
possession, 40
presentience (prediction), 13-14, 45, 51-53, 57
Princeton Engineering Anomalies Research laboratory (PEAR), 30, 65

probability, 6, 8-9, 10-11, 41
  cloud, 47
  statistically improbable events, 52
psilocybin, 22
psyche, 75, 85
  breakdown of, 40, 69,
  components of, 43-44, 87, 97
  porosity of, 49, 58, 59
  See also self
psychedelics, 21-22, 37, 114
  psychedelic experiences, 21- 22, 26, 114, 120-121
  See also consciousness: alterations of, drug-induced
psychism, 49, 51
psychoanalysis 37, 54, 73
psychologists, 39, 49
psychology, 22, 28, 93
psychosynthesis, 43-44, 63
psychotherapy, 73
Puthoff, Hal, 64-65

quantum mechanics, 7-12, 41-42
  backward causation, 11
  and consciousness, 41-42
  decoherence, 42
  delayed-choice experiment, 11
  double slit experiment, 10
  electrons, 10-11, 47
  entanglement, 8-10, 127 n.2, 128 n.2-3
  hidden variables, 8, 42
  Malus' cosine-squared law, 8-9, 128 n.3
  photons, polarized, 8-9, 127 n.2, 128 n.2-3
  superposition of states in, 81
  waves, 10-11, 45
  See also Bell's theorem

random event generator (REG), 5-6, 13-14, 30, 43
reality
  as abstract, 103
  alternate, 22-23, 114
  vs. appearance, 101-103, 104, 107, 109, 111-112
  beliefs about, 16, 19, 21, 23, 24, 32, 45, 64
    personality correlates of, 19
  as code to be deciphered, 58-59
  contemplation of, 18

reality (cont.)
  dimensions of
    other, 23-24, 43-44
    transcendent, 2, 42-43, 56, 59, 68, 94, 105, 118
  fundamental questions about, 1
  ideas about, 10, 23, 24, 31, 35, 39, 59, 68, 118, 120, 123
  identification with, 36
  implicate order of, 43
  interpretation of, 6, 9, 18, 24, 25, 27, 31, 34, 41, 44-45, 87, 116
  knowledge of, 26, 27-28, 36, 88, 93, 119, 126
  materialist view of, 6-7, 15, 41, 44-45
  and mathematics, 99, 115
  and meaning, 17, 22, 116
  mind-like qualities of, 20, 59
  nature of, 9, 18, 20, 24, 25, 27-28, 36, 43, 49, 58, 59, 115, 116, 118-120
  non-physical aspects of, 17, 19, 21, 22, 36, 42-44, 58
  personal aspect of, 119-120
  physical, 43-44, 59, 75
  as that to which one must sacrifice oneself, 110, 123
  subjective understanding of, 25-26, 31, 33
  ultimate, 18, 119, 126
  See also consciousness; feelings: of reality
reasoning. *See* thinking
Reed, Henry, 78-79
REG. *See* random event generator
reliability of personal reports, 23
religion, 17, 32-33, 45
  fundamentalism, 35
  religiosity, 32, 34, 37
    benefits of, 34, 38, 45
    harmful aspects, 35
    as form of intelligence, 36
    and psychopathology, 35, 38
  religious experience, 37, 104
  religious orientations, 32-33, 35, 45
  religious quest, 33, 45, 87
  religious terminology, 2, 32
  and science, 32, 34, 87
  See also spirituality
remote viewing, 64-67, 86
  See also anomalous phenomena

Index

resonance, 6, 14, 20, 30, 45
Roberts, Bernadette, 119

sacrifice. *See* self-sacrifice
Sangster, Lee-Anna, 121
savants, 55-58, 129 n.10
Schatzman, Morton, 77-78, 80
Schwartz, Gary, 60
science, 1, 2, 15-16, 30, 33, 111
   authentic, 15-16, 27-29, 32, 34, 45, 87, 110, 125
   essence of, 27-28, 87
   and existential questions, 34, 125
   inauthentic (*see* scientism)
   interface with spirituality, 2, 117
   politics of, 15, 28, 45
   practice of, 27, 28, 110, 118, 125
   religion, parallel with, 34
   as spiritual practice 87, 89, 93, 111, 117-118, 123, 125
   state-specific science, 118
   transcendent conceptualization of, 30
   world view of, 27-28
   See also empirical studies; knowledge; understanding
scientific community, 1, 15, 30, 118
scientific fact, 6
scientific knowledge, 6, 29-30,
scientific work, 16, 110, 117, 125
scientism, 15, 28, 33-34, 93 (*see also* materialism)
scientists, 1, 2, 5, 14-16, 20, 45, 110-111, 117-118, 125-126
   beliefs of, 20, 45, 125
   as knowledge seekers, 30
   and mathematical yoga, 111
   and religion, 32
   and sacrifice, 110, 125
self, 44, 105, 112
self-determination, 87
   See also authenticity
self-exploration, 18, 31, 32, 45
self-identity, 40, 96, 111
   disidentification with self, 111
   dissociative identity disorder, 40
   self-annihilation, 110-111, 123
   self-identification, 44, 111
self-knowledge, 31
self-mastery, 16
self-sacrifice, 109-111, 117, 123, 126
self-transformation, 2, 18, 23, 34, 39-40, 63-64, 87, 122
   in the context of science, 45, 107, 111, 116, 118, 123, 125
September 11, 2001, 13-14
Shanon, Benny, 114-115
skin-conductance, 12-13
sleep, 23, 76, 78, 85
   falling asleep, 67-68, 69, 73-75, 76, 78, 79-81
Society for Scientific Exploration, 5
spirituality, 1, 17, 32, 35, 36, 119
   benefits of, 38, 119-120
   and consciousness, 17
   and gratitude, 35
   and mathematics, 109
   and psychopathology, 37-41
   significance of, 34
   spiritual aspiration, 1, 2, 49, 87, 106, 110-111, 125
      East vs. West, 106-107, 111
   spiritual entities, 59-60
   spiritual development, 34, 40-41, 125 (*see also* self-transformation)
   spiritual insights, 38, 39, 44
   spiritual intelligence, 36-37, 45
   spiritual practice, 1, 2, 32, 92, 93, 117, 123, 126 (*see also* mathematical yoga)
   spiritual teachings, 49
   See also religion; transcendent states of consciousness
superstitiousness, 18
symbols, 48, 50, 67-68, 77-84, 103, 106
   in dreams, 73, 77-84
      use in dream incubation, 78
   interpretation of, 67-68, 78-84
   See also hypnagogic imagery
synchrony, 6, 7, 8, 10, 13-14, 43

Tart, Charles, 118
telepathy, 19, 75-76
teleportation, 47-49, 59
thinking
   abstract, 101, 102, 107, 110-111, 116-117, 123, 125
   biases in, 56-57
      confirmation bias, 56-57
      truth monitoring function, 56
   conception, 101-102, 107
      semantic nature of concepts, 102-104, 115-116
   critical, 56-57
   deliberate, 61-62, 65-66, 77

thinking (cont.)
  limitations of ordinary, 93
  logical reasoning, 31, 53, 117
  logical structure of, 54, 114-115
  mathematical reasoning, 102, 116, 123
  ordinary, 98, 116
  phenomenology of, 53
  rational, 87
  spontaneous, 61-63, 65-66, 69, 77, 81-82
  as stream of consciousness, 61
  as synthesis of nonconscious and conscious processes, 54
  theoretical, 106-107, 109, 111, 123
  transcendent, 95, 103
  See also consciousness; experiential stream; introception; understanding
trance, 6, 13, 30, 69-71
transcendence, 20, 21, 35, 36, 105, 107, 125-126
  definition of, 92
  nature of, 111
  seeking 92-123
transcendent experiences, 17, 36, 38, 92, 114, 119, 120, 122
  continuum of, 92
  and psychopathology, 38-40
  as superior to ordinary experience, 92
  See also experiences; mystical experiences
transcendent states of consciousness, 34, 37, 88-89
  affective characteristics of, 91-92, 96-98, 119-120, 122-123
  attainment, 94, 97, 100, 105-113, 115-118, 120, 122-123, 125
  endarkenment, 91-92, 98
  enlightenment, 2, 38, 49, 93-98, 109, 112, 117-118, 119-123
  imperience, 96, 115, 123
  movement between ordinary and transcendent states, 94-98, 106-107, 111
  noetic quality of, 36, 49, 93, 97, 120, 122
    as challenge to conventional science, 93
  persistence of, 93, 98
    effects on ordinary consciousness, 96

transcendent states of consciousness (cont.)
  personal aspect of, 119
  scientific approach to, 2, 93
  structure of, 95
    depth, 104
  value of, 119
  See also mystical experiences; noesis
transliminality, 40, 70

Ullman, Montague, 85
understanding
  and authenticity, 33, 63-64, 85,
    definition of authenticity, 87
  depth of, 25, 104, 109
  development of, as spiritual practice, 117-118, 126
  as epistemologically primitive, 31
  as essence of authentic science, 16, 27-28, 30
  need for experimentation with, 123
  extraordinary components of, 88, 104, 117
  as information, 31, 104
  infusion with transcendent insights, 117, 123
  as knowledge, 120
  mathematics, 109-110
  modes of, superior to rational thought, 17, 31, 49
  personal, 25-26, 30, 33, 68, 101, 104
  as personality trait, 19
  as psychologically primitive, 87
  in remote viewing, 65-66
    lack of empirical studies of, 66
  as synthesis, 31, 66-67, 82, 87-88, 122, 125
    of dialectical process, 31
  See also knowledge

Vaughan, Frances, 73
volition
  and authenticity, 63, 66, 87
  as characteristic of deliberate thinking, 61
  in dissociative phenomena, 60, 70
  in dreams, 77
  role in remote viewing, 65-66

volition (cont.)
  role in the induction of lucid dreams, 73
  role in the manifestation of reality, 42-44, 47-48
  and sacrifice, 110

Wallas, Graham, 61
Weyl, Hermann, 106
will. *See* volition
Wolff, Franklin, 1, 2, 93-112, 114-117, 119-120, 122-123, 125, 129 n.13-14
  definition of consciousness, 103
world. *See* physical world
world view, 7, 15-16, 24
  interpretation of world, 18, 27
  materialist, 15, 23, 34, 44-45, 53
  rigidity of, 27-28
Wren-Lewis, John, 91-93, 98, 117, 119, 123, 129 n.12